CAMBRIDGE LIBRARY COLLECTION

Books of enduring scholarly value

Earth Sciences

In the nineteenth century, geology emerged as a distinct academic discipline. It pointed the way towards the theory of evolution, as scientists including Gideon Mantell, Adam Sedgwick, Charles Lyell and Roderick Murchison began to use the evidence of minerals, rock formations and fossils to demonstrate that the earth was older by millions of years than the conventional, Bible-based wisdom had supposed. They argued convincingly that the climate, flora and fauna of the distant past could be deduced from geological evidence. Volcanic activity, the formation of mountains, and the action of glaciers and rivers, tides and ocean currents also became better understood. This series includes landmark publications by pioneers of the modern earth sciences, who advanced the scientific understanding of our planet and the processes by which it is constantly re-shaped.

Geology of Arran and Clydesdale

James Bryce (1806–77) was a Scottish schoolteacher and geologist. His numerous articles on geology earned him a place in the Geological Society of London (1834) and in the Royal Society of Edinburgh (1875). He also campaigned to reform the Scottish universities and for Scottish education to be independent of the English system. In 1855 Bryce conducted a geological survey of Clydesdale and the Isle of Arran for the British Association for the Advancement of Science (BAAS). His findings were published the same year, in this book. Bryce's study records the natural history of the two regions, with descriptions of the geological features encountered on various expeditions in Arran. Bryce also describes the Arran flora, marine fauna, and rare insect life. This book remained the geological authority on Arran and Clydesdale for a long time; the third edition, reissued here, was published in 1865.

Cambridge University Press has long been a pioneer in the reissuing of out-of-print titles from its own backlist, producing digital reprints of books that are still sought after by scholars and students but could not be reprinted economically using traditional technology. The Cambridge Library Collection extends this activity to a wider range of books which are still of importance to researchers and professionals, either for the source material they contain, or as landmarks in the history of their academic discipline.

Drawing from the world-renowned collections in the Cambridge University Library, and guided by the advice of experts in each subject area, Cambridge University Press is using state-of-the-art scanning machines in its own Printing House to capture the content of each book selected for inclusion. The files are processed to give a consistently clear, crisp image, and the books finished to the high quality standard for which the Press is recognised around the world. The latest print-on-demand technology ensures that the books will remain available indefinitely, and that orders for single or multiple copies can quickly be supplied.

The Cambridge Library Collection will bring back to life books of enduring scholarly value (including out-of-copyright works originally issued by other publishers) across a wide range of disciplines in the humanities and social sciences and in science and technology.

Geology of Arran and Clydesdale

With an Account of the Flora and Marine Fauna of Arran

James Bryce

CAMBRIDGE UNIVERSITY PRESS

Cambridge, New York, Melbourne, Madrid, Cape Town,
Singapore, São Paolo, Delhi, Tokyo, Mexico City

Published in the United States of America by Cambridge University Press, New York

www.cambridge.org
Information on this title: www.cambridge.org/9781108038300

This edition first published 1865
This digitally printed version 2011

ISBN 978-1-108-03830-0 Paperback

The original edition of this book contains a number of colour plates, which have been reproduced in black and white. Colour versions of these images can be found online at www.cambridge.org/9781108038300

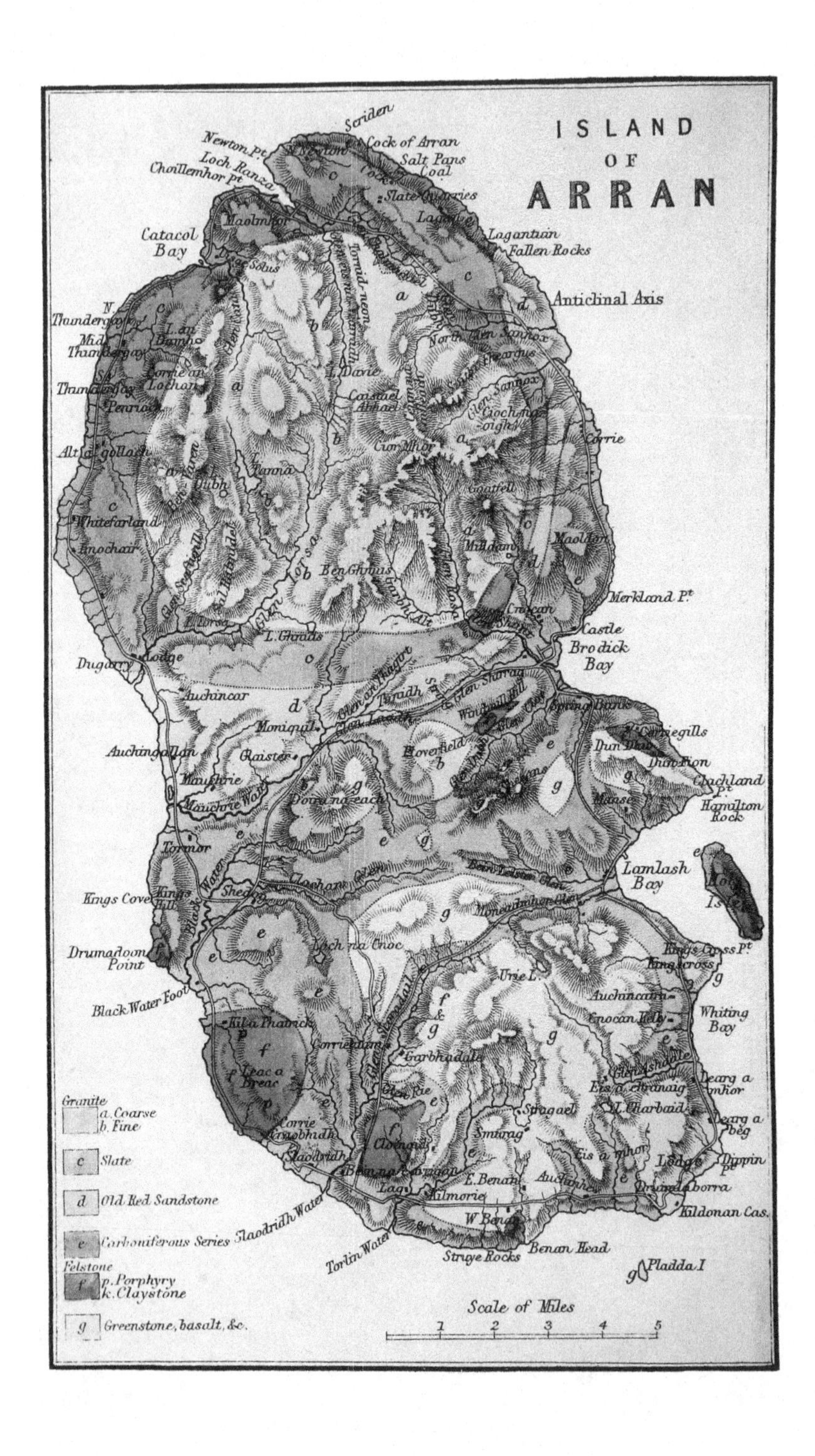
ISLAND
OF
ARRAN
Scriden
Newton Pt
Cock of Arran
Salt Pans
Coal
Loch Ranza
Choillemhor Pt
Slate Quarries
Catacol
Bay
Lagantuin
Fallen Rocks
Anticlinal Axis
North Glen Sannox
Glen Sannox
Corrie
Cior Mhor
Goatfell
Milldam
Maoldon
Merkland Pt
Castle
Brodick
Bay
L. Davie
Tanna
Whitefarland
Enochair
Ben Ghnus
L. Iorsa
L. Ghaids
Dugarry
Lodge
Auchincar
Moniquil
Glaister
Auchingallan
Mauchrie
Mauchrie Water
Spring Bank
Corriegills
Dun Dhu
Clachland Pt
Hamilton Rock
Lamlash
Bay
Holy Isle
Tormor
Kings Cove
Kings Hill
Black Water
Drumadoon
Point
Black Water Foot
Loch na Cnoc
Kings Cross Pt
Kingscross
Urie L.
Auchincairn
Whiting
Bay
Kilpatrick
Corriegillan
Garbhadale
Leac a Breac
Dearg a mhor
Dearg a beg
Stragael
Smirag
Dippin Pt
Lodge
Drumlaborra
Kildonan Cas.
E. Benan
Lag
Kilmorie
W. Benan
Benan Head
Struye Rocks
Torlin Water
Pladda I
Granite
a. Coarse
b. Fine
c Slate
d Old Red Sandstone
e Carboniferous Series
Felstone
f p. Porphyry
k. Claystone
g Greenstone, basalt, &c.
Scale of Miles
1 2 3 4 5

GEOLOGY

OF

ARRAN AND CLYDESDALE

WITH AN ACCOUNT OF

THE FLORA AND MARINE FAUNA OF ARRAN

NOTES ON THE RARER INSECTS AND NOTICES OF THE SCENERY AND ANTIQUITIES

BY

JAMES BRYCE, M.A., LL.D., F.G.SS. L. & I.

THIRD EDITION

GLASGOW
THOMAS MURRAY AND SON
EDINBURGH, J. MENZIES
LONDON, JAMES NISBET AND CO.
1865

PREFACE

THE following work originated in a request of the Local Committee of the British Association, on the occasion of the last meeting of that body in Glasgow. The Committee was desirous that such an account of the geological structure of the neighbourhood should be prepared as would serve for a guide to the Geologists of the Association in exploring Clydesdale and the shores of the frith. To this request the Author willingly acceded; and to the sketch then printed he appended a notice of some peculiarities in the structure of Arran.

Finding sometime afterwards that many persons wished for a more detailed account of the geology of these districts than was given in this sketch, the Author republished it in a second and enlarged edition. The part of that edition which related to Arran was entirely new, and was founded upon the observations made during several successive summers among its romantic glens and mountains.

As Arran is a highly interesting field in the departments of botany and marine zoology, the Author considered that the study of these branches would be promoted, and the wants of the student met, by combining an account of the Flora and marine Fauna with that of the geology. At his request, the Rev. Dr Miles, well known for his researches in the Clyde, conducted under the auspices of the British Association, kindly drew up an account of the more remarkable objects of the marine Fauna. The complete list of species which Dr Miles has given, brought down to the time of his leaving Glasgow for Malta, will prove welcome and useful to naturalists. Dr Carpenter, who has resided in Arran for several summers since Dr Miles left this country, has been so good as to supply an interesting addition to his account. The account of the Flora of Arran has been drawn up by the Author's eldest son, Mr James Bryce, Fellow of Oriel College, Oxford. To Professor Balfour of Edinburgh the Author's best thanks are due for the permission kindly granted to make use of his catalogue of Arran plants.

The Author has also to acknowledge his obligations to a distinguished Entomologist, H. T. Stainton Esq., of Mountsfield near Lewisham, for the notes which he has most kindly supplied on the rarer insects of Arran.

To Colonel Sir Henry James, R.E., and Lieut.-Colonel Bayly, R.E., he has also to express his best thanks for their kindness in supplying him with the heights of the Arran mountains, ascertained in the Ordnance Survey lately made, but not yet published.

In examining the lower members of the coal series in the Campsie district the Author was assisted by Mr John Young, now of the Hunterian Museum, Glasgow; and many facts stated regarding them were supplied by him.

The present Edition has been made larger and more complete than its predecessor by adding several new Excursions, an account of the Arctic-shell beds lately discovered in Arran, and notices of the excavations made within the Stone Circles; as well as by introducing much new matter relating to the antiquities, history, and traditions of the island. Thus expanded and carefully corrected throughout, the work may serve as a general Guide to Arran.

On the portion relating to Clydesdale improvements not less important have been made. Since the second edition was published, so much has been done among the shelly deposits, which Mr Smith of Jordanhill established some years ago as a distinct group, that it has been necessary to write a new and much fuller account of them. In drawing up this account the Author has been kindly assisted by the Rev. H. W. Crosskey of Glasgow, one of the most zealous and successful explorers of these beds since Mr Smith ceased to work actively among them. It is hoped that the view now presented of their physical sequence and fossil contents will be found the most complete and satisfactory which has yet been given. It is hoped also that additional interest and value will be given to the sketch of the carboniferous formations by the tabular view of the economic strata now inserted. It has been kindly drawn up for this work by Mr William Moore, civil and mining engineer, Glasgow, and exhibits these strata as they now exist in our coal-fields, with the estimated quantity of each still remaining unworked.

GLASGOW, 15*th September*, 1864.

CONTENTS

GEOLOGY OF ARRAN.

Excursions in Arran.

THE FLORA OF ARRAN.

MARINE ZOOLOGY OF ARRAN.

ENTOMOLOGY OF ARRAN.

GEOLOGY OF CLYDESDALE.

GEOLOGY OF ARRAN

1. THE Island of Arran is twenty miles and a half long from N.N.W. to S.S.E.; ten and a half miles broad, and with Holy Isle and Pladda includes an area of 103,953 acres.—The number of rock formations, sedimentary and plutonic, which are found within this limited space, is truly remarkable, perhaps unparalleled in any tract of like extent on the surface of the globe; while the varied phenomena which they present in their mutual contacts and general relations to one another, are of the highest import in theoretical geology. The variety indeed is so great, and the interest so lively and pleasing, which an examination of the structure of the island and its charming scenery excite, that, as Professor Phillips has remarked, every geologist who visits Arran is tempted to write about it, and finds something to add to what has been already put on record. For the student there cannot be a finer field; the primary azoic rocks, the metamorphic slates, the lower palæozoic strata, the newer erupted rocks, and phenomena of glacial action, may all be examined by him in easy excursions of a few days; and the exposition of the strata is so complete in the rugged mountains, deep, precipitous glens, and unbroken sea coast sections, that the island may truly be called a grand museum, arranged for his instruction by the hand of Nature.

Physical Features

2. A line running from the north angle of Brodick bay almost due west to Iorsa waterfoot, divides the island into two

nearly equal portions, strikingly different in their geological structure and in their outward features. The northern half consists of a mass of peaked and rugged mountains, intersected by deep and wild glens, which diverge from a common centre, and open seaward on a narrow belt of low land. This belt forms a terrace marking the ancient sea-level, and is bounded inland by cliffs pierced with caves, and otherwise sea-worn. The coast road is carried along it from Brodick bay to the mouth of South Sannox water, and again from Loch Ranza to Dugarry at the mouth of Iorsa water, and affords throughout views of surpassing beauty. This terrace is a striking feature of the Clyde shores everywhere, and will be more fully described when we come to notice the tertiary deposits.

The Arran mountains are naturally divided into three separate ridges, which may be named after their most conspicuous summits—the Cior-Mhor, Goatfell, and Ben-Varen groups. Of these, by far the most considerable is the first, which forms a long, irregular, narrow, and jagged ridge, extending from Ben-Ghnuis on the south to Suithi-Fergus on the north-east. Cior-Mhor stands near its centre at a point where a salient angle in the ridge closely approaches to the Goatfell group on the east. Its connection to this group by a cross ridge or col about 1,000 feet in height, which separates the heads of Glen Rosa and Glen Sannox, and its position near the middle of the range, constitute the prominent peak of Cior-Mhor the geographical centre of the whole north-eastern mass of mountains. Its height is 2,700 feet; but it is exceeded by Caistael-Abhael, and perhaps another summit. The whole ridge has great persistent altitude, no point descending below 1,600 feet, and there being at least six summits not less than 2,000 feet. Fronting the concavity of this arch-like ridge on the east, is an assemblage of closely connected mountains dependent upon Goatfell. The most northerly summit is the lofty conical peak of Cioch-na-Oich, or the Maiden's Breast, guarding the south-east angle of Glen Sannox; on the south the group terminates in the bold precipice called Glen Shant rock, at the entrance of Glen Rosa. Both groups front the interior glens in tremendous precipices, while they descend with less abruptness to Glen Iorsa on one side, and the sea on the other. To the west of the geographical axis of the island lies the Ben-Varen range, with some connected and lower heights east of it, the entire group being separated from the other mountains by Glen Iorsa and Glen Eais-na-Vearraid, running respectively S.S.W. and N.N.W. through the length of this half of the island. These glens have a common watershed in Loch-an-Deavie, a small mountain lake or tarn, which, when it stands at a high level, as in winter and in wet summers, discharges its waters at both ends. Another small lake adjoining, at a higher level, likewise discharges, when full of water, north-west into the head of Glen Catacol, and south-east into Glen Iorsa; but it is a mistake to suppose that Loch Tanna, the largest lake in the island, has this singular position; the ground rises suddenly north of it, and all the water passes off into the Iorsa. These Arran lakes illustrate an arrangement which occurs on a large scale in Canada and Norway, where lakes extend across the entire watershed between two seas.

The bleak uplands between Ben-Ghnuis and Ben-Varen are finely varied by several heights, of which the most remarkable is the prominent ridge of Sal-Halmidel, an outlier of the latter range. It divides Glen Iorsa from Glen Scaftigill, and is a conspicuous and picturesque object from all the south-western portions of the island. To complete our sketch of the northern section, it is only necessary to notice the high ridge between the sea on the north, and north Glen Sannox and Glen Chalmidel on the south. In geological formation it differs from the others, but is connected with the main range, which we have called the Cior-Mhor group, by means of a ridge at the watershed between the two glens which bound it. This ridge joins on to the eastern flanks of Tornidneon, an abrupt, massive, but not lofty mountain, overlooking Loch Ranza on the south, and forming the termination of a long but not generally high ridge which sweeps round the east side of Glen Eais-na-Vearraid, and runs in upon the north side of Caistael-Abhael near the watershed at Loch-an-Deavie. This northern range terminates in bold heights forming the east side of Loch Ranza; its western boundary is formed by the nearly precipitous sides of Meal-Mhor, the most northerly mountain of the Ben-Varen group. Thus at the northern apex of the island, the principal ranges closely approach one another, their terminal portions forming the lofty abrupt framework to the secluded inlet of Loch Ranza, whose many picturesque features excite the admiration of every visitor.

Various estimates have been formed from time to time of the heights of the Arran mountains; all of them, including some of our own made by trigonometrical measurement, pretty near the truth. They have lately been determined, in the conduct of the Ordnance Survey, with that exactness which marks all the operations of the Royal Engineers. The following are the principal heights:—Goatfell, 2,866 feet; Caistael-Abhael, 2,817; the Ceims, 2,706; north top of Goatfell, 2,628; Ben-Ghnuis, 2,597; Ben-Varen, south summit, 2,342; north summit, 2,310; Cioch-na-Oich, 2,687; highest point of the slate ridge, east of Loch Ranza, 1,097 feet; Holy Isle, 1,020; Benan Head, 457; West Benan, 523.

3. The southern half of the island consists of a rolling table-land, bleak and unpicturesque inland, but breaking rapidly down seaward into a coast border of great romantic beauty. The general elevation is from 500 to 800 feet; and the irregular ridges which traverse it, most usually in a direction nearly east and west, rise to about 1,100 or 1,400 feet. These separate the various glens and river-courses, whose origins lie near the central line or axis of the island, but often interlace with one another, so that a stream issuing westward has its source nearer the east coast than the west, and *vice versa*. The views of the northern mountains, from these uplands, are very grand, especially when they are seen in the early summer twilight, their dark jagged peaks projected against a background of

sky, still lit up with brilliant hues from the departed sun. Their aspect from one such point of view has been given with noble effect by a living artist.—The terrace border, so conspicuous around the estuary of the Clyde, is marked in the southern section of the island with less continuity than round the northern part, owing to the nature of the rocks, and their advance in many places upon the sea-line in mural precipices. Still, however, it is sufficiently distinct in many parts, as about Corriegills, Whiting Bay, portions of the south coast, and towards King's Cove on the west, where its salient and re-entrant angles, in their bold points and noble sweeps, afford some of the finest scenes of quiet beauty to be met with in Scotland.

General Outline of the Structure

4. The remarkable geological structure of Arran, and the striking physical features which give such a charm to its scenery, are alike due to a single peculiarity—the *abnormal position* of its granite nucleus. The other granite tracts of the north-west of Scotland lie amid primary slates, which are symmetrically disposed on opposite sides of it, the granite ridge forming an anticlinal axis. This is the case throughout the Grampians, where gneiss and mica slate widely encircle the various outbursts of the granitic rocks; the silurian slates of the south of Scotland envelope in the same manner the granitic bosses which rise amid them. But in Arran there is no such symmetrical arrangement; granite does not form a mineral axis in reference to the slate rocks. It has been protruded *close to the outer border* of the two upper slates, so as to come into contact on one side with the newer sedimentary strata. Its position is thus very different from that of the Grampian granite, being far removed from the axis of the old crystalline slates, and associating it with sedimentary strata of a much later age. A broad zone of slate rock traverses Scotland diagonally from sea to sea, intersecting the line of the Clyde shores, and crossing the islands of Bute and Arran. It is usually divisible into three distinct bands—a lower micaceous,

passing into gneiss, and stretching inwards towards the Grampian axis; a middle dark-coloured clay slate, and an upper green or chlorite slate, the two latter forming what is called the clay slate series. The newest of the two upper slates is not found in Arran, and is probably either thrown out westward beyond the line of bearing of the second slate, or is so altered by the near proximity of the granite as to be undistinguishable from the middle or dark-coloured slate, through which the granite has been protruded. On the east side of the granite nucleus, above Corrie, this slate band is extremely narrow, probably only a few yards thick in many places; on the western side it is much broader, but the underlying mica slate does not appear, and is not found till we pass into the opposite peninsula of Cantire, where we meet with it on arriving at the line of bearing which it preserves, as already remarked, in crossing the country from north-east to south-west. To this succeed the gneiss tracts, which, as we advance north-west, keep with us till we reach the granite axis of the Grampians. Along this axis, however, it is now known that the granite does not form an unbroken continuous ridge or line of heights. It occurs rather in independent, elliptic-shaped masses, its place on the axis being often taken by mountains of gneiss, porphyry, or quartz rock.

Such, then, is the singular abnormal position of the Arran granite, which gives to this island all its peculiarities, both as regards its geographical features, and its geological structure. The protrusion of so large a body of igneous rock by plutonic fires along the line of junction of the older slates and the secondary formations, and its elevation to a great height in a space so limited, might naturally be expected to produce phenomena of varied interest, such as have been alluded to in the opening paragraph, (Art. 1.)

The Slates

5. The granite nucleus occupies by far the greater portion of the northern half of the island. The three mountain groups

already described, with the glens and valleys penetrating and dividing them, consist entirely of this rock. It is remarkable, however, that at no point does the granite reach the sea coast. In its elevation from the plutonic depths, it bore up with it a narrow band or framework of clay slate, of the second or dark-coloured variety, which completely encircles the nucleus. The structure of this portion of the island is shown in the cut annexed, No. 1, representing an ideal east and west section

No. 1.

(a) Coarse-grained granite; (b) fine-grained granite; (cc) slate; (d) old red sandstone; (e) carboniferous strata.

from Corrie to White-Farland. From the mouth of the Iorsa water at Dugarry, round the west side of the island by Imochair, Thundergay, and Catacol, to Loch Ranza, this slate occupies the coast, and forms a belt of considerable but varying breadth. Its junction with the granite is seen in almost every stream, and in many points along the western slopes of the lower hills. It extends all round the precipitous sides of the Loch Ranza valley, and to a short distance east of Newton point, which forms the north-east angle of the loch, as far as a small stream called Alt-Mhor. It here retires from the shore, and forms the high northern ridge already mentioned, the coast from this point eastwards, and then southwards, being occupied partly by old red sandstone and partly by coal sandstones, beds of carboniferous limestone and of coal and coal shale being interposed amid the beds of the latter. The band of slate is of considerable breadth in this northern ridge, but narrows very much on approaching the opening of Glen Sannox; along the hill-sides southward, from the base of Cioch-na-Oich to the slopes over Corrie, the breadth varies from 25 to 40 or 50 yards. Farther south, as it sweeps round south-west between Maoldon and the base of Goatfell, it gradually widens

—presents bold precipices on both sides of the lower part of Glen Rosa, and along the south border of the granite nucleus between Loch Ghnuis, Dugarry and Imochair, attains its greatest development. Here the thickness of the slate band between the granite and old red sandstone cannot be less than 2,000 or 3,000 feet; whereas above Corrie, in the bed of a stream, north of the White water, we noticed a spot where it is, most probably, not more than ten yards, so close to the outer margin has the remarkable outburst of granite taken place. It has indeed been asserted by one writer, and repeated by others, that the continuity here is entirely broken, and that the granite is in contact with the old red sandstone. This we consider is decidedly a mistake. In frequent summer rambles along these romantic hill-sides, we have examined carefully every open section, and have never seen such a contact. In the spot referred to, the rock is a hard, bluish-gray, flinty slate, very like a hard sandstone when taken wet from the burn. The strata here come against the granite end-on, or nearly at right angles. The prevailing dip of the slate over the whole district is the same as that which it maintains in adjoining tracts beyond the limits of Arran, namely, south or south-east at a high angle; it is not arranged in mantle-shaped strata around the granite nucleus, but is inclined towards it in some places, and off from it in others. Thus, along the north side of the nucleus from North Sannox to Catacol, the dip is south or towards the granite, at angles varying from 65° to 75°. This is of course unconformable to the granite centre; but a like dip and inclination on the south side give conformability; and here, accordingly, from Maoldon by Glen Rosa and Dugarry, and perhaps even as far as Thundergay, the slate is seen to recline against the sides of the granite mountains in a kind of mantling stratification. In many spots there is great contortion and irregularity, indicating the operation of violent forces, attendant on the upheaval of the granite. Reference will again be made to several of these cases, as well as to local variations in the mineral character of the slate.

Granites of the Nucleus

6. Within the granite tract, whose limits are defined by the slate as above described, there occur two distinct varieties of the rock—a coarse and a fine-grained, occupying separate tracts. Dr M'Culloch was the first to describe them as distinct. Mr Ramsay traced their limits in a general way; in our former edition we extended and more exactly defined these limits. In April of the present year we discovered that the fine variety is not isolated, as was supposed, within a boundary of the coarse kind; but that it comes into contact with the slate along the south side of the nucleus. Thence it stretches north across the watershed into Glen Eais-na-Vearraid, and occupies the space between the Ben-Varen and Cior-Mhor ranges, rising high on the east front of Ben-Varen and west front of Ben-Ghnuis, and as far as the passes into Glen Rosa and Glen Sannox, on either side of Cior-Mhor. The relation of the rocks on a north and south section, from Loch Ranza to Mauchrie-water, is shown in the annexed cut:—

No. 2.

(*a*) *Coarse-grained granite;* (*b*) *fine-grained granite;* (*cc*) *slate;* (*d*) *old red sandstone;* (*e*) *granite of Craig-Dhu.*

The coarse variety covers the rest of the district, and forms the tops of all the mountains. The Goatfell group is composed entirely of it; as are also the whole east front and all the summits of the Cior-Mhor range from Ben-Ghnuis to Suithi-Fergus. The two kinds are very distinct from one another, though the component minerals are the same. Both consist of quartz, felspar, and mica, the two first being in nearly equal proportions, the last in less quantity than either. In neither kind is the mica replaced by hornblende, so as to form the variety called syenite. This rock is indeed found in Arran, but not within the district called the granite nucleus.

There are, however, several varieties of both kinds; but these are merely dependent upon slight changes in the colour or size of the constituent crystalline grains. The fine variety of the interior has often an arenaceous aspect and sandy feel, from the minuteness of the grains. The distinctness of these two varieties of granite in their mineral aspect, and the fact that they do not intermingle, but occupy separate tracts, led Mr Ramsay to conclude that they were of different ages, and that this fine-grained variety was the newest rock in the whole island. He rests this conclusion on the non-existence of whin dikes in the fine granite, and on what he seems to state with the confidence of personal observation, namely, that "in cutting through the coarse granite, these dikes frequently approach the borders where it joins the fine variety, and are then invariably cut sharply off by the fine granite, when they approach it in the coarser kind" (p. 65). We have never noticed any instance in which a dike is thus cut sharply off, nor does Mr Ramsay refer to any spot where such a case occurs; and it is possible he may refer merely to the non-existence of dikes in the fine granite, while they are often met with in the coarse. He states his view with some diffidence, and with the qualification that the tract of fine-grained granite has not been carefully investigated with respect to these appearances. This examination was carefully made by us; and the result was the discovery of several trap dikes, ranging between north and north-west across the tract of fine-grained granite; but the intersection of these with the coarse-grained variety could not be observed, nor has such intersection been anywhere noticed by us. It is quite possible, however, that there may be one system of dikes in the coarse granite cut off by the fine, and another system in the fine itself. If the intersection does indeed occur in the manner mentioned in the above quotation, the conclusion will be forced upon us that the fine granite is not only newer than the coarse, but newer also than the dikes in this coarse variety. There would thus be granite of two ages, and trap of two ages, in this central district. We must, however, express our belief that the discovery of trap dikes traversing the fine granite for considerable distances renders it extremely probable that they are continued both ways into the coarse variety; and that, while the two granites may be of different ages, the dikes are of one age and posterior to both granites. We shall have occasion hereafter to refer to the age of these two granites. The veins of fine granite—which are so frequently found traversing the coarse granite of Goatfell, Ben-Ghnuis, and other localities—are certainly cotemporaneous, formed by chemical affinities or electric forces acting in some peculiar way during the consolidation of the rock, and not injected from below at a later date. They are abundant along the crest of the Goatfell range, which is separated from the tract of fine granite by Glen Rosa and the ridge of the Ceims, and it is inconceivable that, from the depth at which the fine variety must exist beneath the coarse, if it do actually impenetrate the coarse beneath Glen Rosa, veins so fine should pass continuously to so great a height. Besides, both veins and elliptic or globular masses of such fine granite are often found quite isolated in the large-grained variety, even when far from any possible connection with the fine-grained granite. We have seen them indeed in the granites of various districts, both of Scotland and other countries, where the rock is all of one structure, and there is no evidence, such as exists in the case of Arran, for a difference of age.

The Old Red Sandstone

7. The encircling band of clay slate is succeeded on the east and south by a band of old red sandstone, which, like the slate band, is of irregular breadth. It begins to overlie the slate at the Fallen Rocks on the north-east coast, and occupies the shore thence to the march of Achab farm, half a mile north of Corrie. Here it retires inland, the carboniferous formations taking its place on the shore; crosses in a narrow band to the west of Maoldon, and stretches thence

continuously westward, around the border of the slate to the mouth of Mauchrie water. Between this point and Dugarry, near Iorsa waterfoot, it attains its greatest breadth. The breadth is also considerable from the Fallen Rocks to a point a little south of the base of Cioch-na-Oich. A line from the north side of Brodick Bay to Dugarry very nearly marks out its line of junction with the slate. In structure it varies from a fine-grained red or dark brown sandstone to a coarse conglomerate, in which the fragments are more than a foot in diameter.

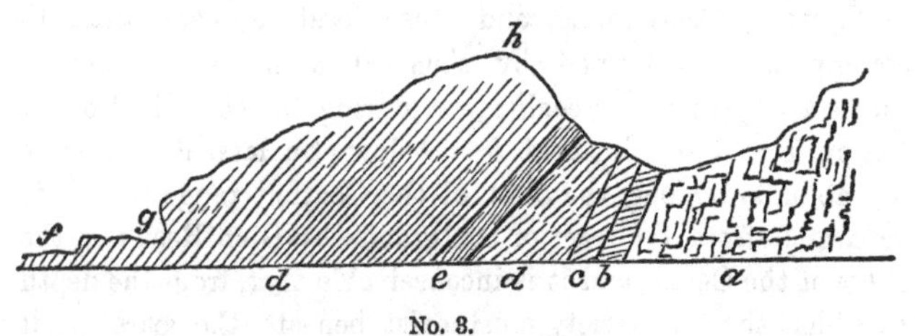

No. 3.

(a) Granite of Goatfell; (b) slate; (c) old red sandstone; (dd) coal sandstone; (e) band of limestone; (f) tideway; (g) terrace and old sea cliff; (h) top of Maoldon.

The coarse and fine strata do not follow any particular order, but alternate throughout the formation, indicating the operation of powerful currents, and intermediate periods of repose.

It may here be merely noticed that the sandstones of the coast have the edges of the strata turned up towards the central granite, and that north of the mouth of North Sannox water, the dip is towards the north-east and north, while south of this point it is south-east and south. Beds of limestone occur subordinate to the old red sandstone at the march of Achab farm, and in a few other places. This limestone is of concretionary structure, without fossils, and similar to the cornstones of England—members of the Old-Red system.

The Carboniferous Series

8. The southern half of the island, southward of the old red sandstone boundary above defined—that is, from the String road and valley of Mauchrie water to the South End—

is composed of several members of the carboniferous series, broken through and overlaid by various igneous rocks, chiefly those of the greenstone type. The prevailing rock and substratum of the whole southern plateau is red sandstone, varying from a fine compact structure to that of a coarse conglomerate. A narrow band of this sandstone extends also along the eastern shore northwards to Corrie; and again from the Fallen Rocks north-west to the Scriden at the northern extremity near the entrance of Loch Ranza. Subordinate to this sandstone are beds of limestone, abounding in fossils of true carboniferous types, beds of shale and coal, in which are found fossil plants and shells, such as characterise these strata in the basin of the Clyde. The structure, indeed, is quite analogous to that of the Clyde basin. Limestone does not occur at the base of the system, nor does it occupy any determinate place in it, but is found throughout the whole series of beds in repeated alternation with the sandstone and shales. These alternations are seen in section on the sea shore along the east border, on the hill-side between Corrie and Brodick, along the high grounds on both sides of Glen Cloy, in Clachan Glen, Glen Scorodale, and other water channels issuing westward; and in the Alaster water above Lamlash they descend almost to the bottom of the glen. On this ground we refer the whole of the strata of the southern division of the island to the carboniferous series. The alternations in question are not seen, it is true, on the Corriegills, Whiting Bay, or Kildonan shores, nor do they appear among the sandstones of the South End; but such massive strata of red sandstone as appear along the east and south shores, are common in the carboniferous system, and form indeed everywhere its prevailing member. They are conformable to the strata inland, in perfect sequence with them, of the same mineral structure, and without any fossils of New-Red types. We do not, therefore, hesitate to refer them all to the carboniferous system; and we are persuaded that the distinguished geologists who placed them in the New-Red series will be the first to agree in this view.

9. The fragments imbedded in the conglomerate, both of this age and of the Old-Red system, are mica slate from a distant source, the ordinary slate adjoining, and quartz of two varieties, of which one is the common white quartz, forming veins in the slate, and the other a peculiar resinous quartz, of a cinnamon colour, not found in Arran, and of which we know no locality, except the eastern valleys of the Ben Nevis group of mountains, south of Bridge of Roy; but beds of this peculiar kind of quartz may undoubtedly occur among the primary slates of Cowal and Cantire, or, it may have some more distant source among the Grampians. It is clearly this north-western region of Argyle which has furnished the fragments imbedded in the conglomerate. The resinous variety is rounded and polished, while the other is often angular or slightly rounded. The former is exactly the same as fragments which occur abundantly and of large size in the conglomerates of the east coast of Antrim, but not known *in situ* anywhere in the north of Ireland. It is remarkable that fragments of granite do not occur in the conglomerate; none of the Arran sandstones have as yet yielded a piece of this rock, except in a single instance, where its presence may be otherwise explained—see Art. 12. The conclusion to be drawn manifestly is—that when these sandstones and conglomerates were in process of formation by the wearing down of the slate rocks and the transport of the fragments by water, the granite of the interior was not exposed to disintegrating causes, but remained as yet in hypogene depths, protected most probably by the enveloping slate rocks. Facts to be stated farther on will throw light on this curious subject.

The Outlying Granites

10. Perhaps the most remarkable feature in the geology of Arran, is one made known by its recent explorers, and of which the British Isles offer, we believe, only one other ex-

ample. We refer to two outbursts of granite amid the sandstones of the southern division of the island, noticed in the last three Articles. One of these was discovered by Mr Ramsay in 1837; but the first description of it was published in 1839 by M. Necker, who named the district Ploverfield. The other was discovered by the writer of these notices in the summer of 1855, and described in the autumn of that year at the Glasgow meeting of the British Association. Both tracts occur on that side of the sandstone district which is nearest to the "granite nucleus;" and the occurrence of the rock here is thus intimately related to the outburst of the central granite, in its abnormal position—close to the outer border of the upper slate, and to the base of the sandstone formations—already pointed out as the great leading peculiarity of Arran. These tracts are represented in the annexed ideal section from east to west between Corriegills and the valley of Mauchrie water, the horizontal extent of the intervening sandstone being much contracted:—

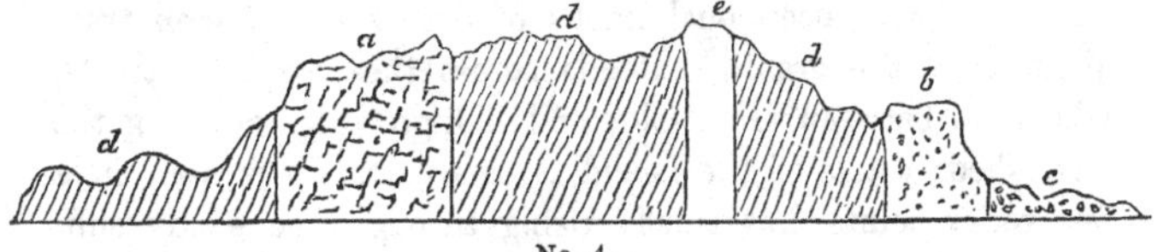

No. 4.

(a) Granite of Ploverfield; (b) new granite tract at Craig-Dhu; (c) old red sandstone of Mauchrie water; (ddd) sandstone with beds of limestone, the whole of carboniferous age; (e) eruptive rocks of Doir-nan-Each, the highest hill in the west district of the south section of the island, chiefly porphyry and highly hornblendic basalts.

(a)—The Ploverfield Granite

11. The Ploverfield granite occurs amid strata which undoubtedly belong to the carboniferous system. The tract is situated on the west side of Glen Cloy, and that branch of it called Glen Dhu. The hills here are the highest portion of a long ridge running up south-west from the plain of Brodick, and dividing Glen Cloy from Glen Shirag. At the origin of this ridge are beds of carboniferous limestone with fossils

under the north wall of the enclosure of Brodick church. On the upper slope of the ridge, beds of limestone and shale also occur in the sandstone, not far from the north base of the Windmill hill—a high, elongated, narrow ridge, steeply overhanging Glen Dhu, and composed of quartziferous porphyry. On its south-west base the granite first appears, separated, in some places at least, from the porphyry by a narrow band of altered sandstone. Thence, keeping at a high level, it extends along the hill slopes at the head of Glen Dhu, and terminates southwards against a ledge or low cliff of syenitic greenstone, a little west of the top of the Blackhill—a high, bold point separating Glen Dhu from the upper part of Glen Cloy. The top and front of this hill toward both glens are composed of altered sandstone, resembling quartz-rock, the change of structure being, no doubt, due to the proximity of this syenitic greenstone. The ledge or cliff seems to be the northern front of a large dike ranging between west and north, and cutting off the granite. To the south of it, the hills are composed of sandstone, with occasional knolls of overlying common trap. Westwards, the granite extends across the moorland tracts, which lie between Glen Cloy and a high ridge forming the watershed or axis of the island at this part. It is seen in a few rocky points, but chiefly rising through the sandy substratum of the peat, where this has been worn away, or in the beds of the small streams. As we advance westwards, it becomes gradually intermingled with syenite and porphyry; but the relation of the three rocks cannot be made out, owing to the nature of the ground. They appear to alternate, the granite diminishing in quantity toward the west. On the summit of the highest ridge or watershed, loose pieces of granite strew the surface, mixed with syenite and porphyry. Bosses of the latter rise through the broken-up masses of the sandstone, forming the main body of this ridge. The surface is also strewn with masses of pitchstone, and of the coarse-grained granite of the northern mountains. We must thus be content with a hypothetical boundary for the Ploverfield granite on the west. Its southern boundary does not reach so far as a line joining the heads of Clachan Glen and Glen Cloy, sandstone being seen continuously between them. On the north side its boundary runs from the west base of the Windmill Hill to a point considerably to the west of the summit level of the String road. Here a high ridge on the south side of the glen, and west of the watershed, presents granite in its upper part; the lower portion, on the level of the watershed, is composed of sandstone; but the junction is not seen. These limits mark out an area much larger than that hitherto assigned to the granite. The weathered surface of the granite is generally white; but the prevailing colour in a fresh fracture is red. It is fine-grained, with little mica in proportion to the quartz and felspar, and is very similar to the fine-grained variety of the interior. Drusy cavities often occur, lined with crystals of quartz and felspar. Mr Ramsay states (*Guide to Arran*, p. 12), that "It sends forth veins into the adjacent sandstone; while specimens of the sandstone, much altered by the effects of intense heat, may even be found enclosed in the granite." We have not noticed these cases of intrusion; nor are the precise spots mentioned. We have, however, seen such in connection with the porphyry of Windmill Hill, to which the granite succeeds on the west, and also in connection with the new granite tract to be next noticed.

(*b.*)—*The Craig-Dhu Granite*

12. The other granite tract lies on the south side of the Shiskin road, nearly opposite the farm-house of Glaister. Here the hill, whose base is skirted by the road all the way down from the "String," overhangs the valley of Mauchrie water in a steep cliff called Craigmore, Craig-Dhu, or The Corby's Rock. This cliff is the outer edge of a small plateau or table-land, cut off from the higher ground behind, towards Doir-nan-Each, by a deep hollow which completely isolates it. The summit is 700 to 800 feet above the valley, and is more than a quarter of a mile long, and from 250 to 300 yards

broad. It descends steeply towards Shiskin on the south-west, and slopes gradually north-east towards Glen Leogh. The summit and sides of this plateau are formed of fine-grained granite, very similar to that of Ploverfield. The base of the cliff towards Mauchrie water is covered by a long talus of granite blocks and smaller fragments, reaching to within 50 or 100 yards of the road, and appearing even at that distance of a very different aspect from fallen masses of sandstone.—See last cut, *b*, *c*.

The granite here seems to rise either through the old red sandstone, or at the junction of this rock with the carboniferous strata. The granite is nowhere seen *in situ* at a low level; the talus before mentioned obscures the rocks along the base of the hill; and the ground by the roadside, and along the valley, is deeply covered with alluvium. At some spots, however, rocks are seen apparently *in situ*. They are not of a very marked character, but seem to be chiefly sandstone of the old red formation, greenstones and trap porphyries, the sandstone having assumed a subcrystalline metamorphic aspect from the intermixture. But at a high level on the west, south, and east sides of the plateau, the granite is seen to rise through a coarse conglomerate; and numerous contacts are observable. These are highly interesting, and clearly indicate the intrusion of the granite subsequently to the formation of the conglomerate. The base of this conglomerate is a coarse sand, and the imbedded fragments sandstone, quartz, and granite. The base is highly indurated, and assumes a porphyritic structure; the sandstone is rendered crystalline, and the quartz has been fused, and reconsolidated into a substance resembling porcellanite. The fragments of granite are of an elliptic form, less rounded than the quartz, and are exactly like the adjoining mass of granite in structure and component parts. Whence have these granite fragments been derived? From the body of fine granite among the northern mountains, or from the adjoining mass itself? Mineral structure does not enable us to determine—the two rocks are so similar. If from the former source, then we must conclude that the granite of the interior was elevated so as to be exposed to disintegrating causes while the conglomerate was forming; in which case granite fragments ought to occur abundantly in the sandstone conglomerates; but this is not found anywhere in Arran—a fact noticed by all observers. Even here the fragments occur only in close proximity to the granite itself. Must we not then rather suppose that pieces of the granite adjoining, when this rock was erupted in a fluid or semi-fluid state, were injected among the outer strata of the conglomerate, also fused by the contact, and so became imbedded in these strata only?

Age of the Granites

13. Granite, then, occurs in Arran, in three disconnected tracts; amid slates in the northern mountains, in the old red sandstone of Craig-Dhu, or at the junction of this rock with the carboniferous strata, and also amidst these carboniferous strata at Ploverfield. In each of these positions it is clearly posterior to the rock which encircles it; for it is intruded among these sedimentary deposits, and produces a marked alteration upon them along the planes of contact. We have, therefore, now to consider the question of age. Are the three granites of three distinct ages corresponding with those of the strata among which they intrude? or were they erupted simultaneously, so as to pierce through the three formations during one and the same period of disturbance? In other words—and this view narrows the question—since the Ploverfield granite is clearly of later origin than the sandstones of Windmill Hill, and the shell limestones subordinate to them, were the granites of Goatfell and Craig-Dhu erupted at the same time with it? or does their injection among the strata, and elevation to the day, date back to an earlier period? The close proximity of the Craig-Dhu granite to the border of the carboniferous formations, if it be not actually enclosed in these, evidently points to an identity of age with that of Ploverfield, and renders their simultaneous eruption extremely probable.

How then is the Goatfell granite related to the old red and carboniferous formations? It has been long established, and is well known that it everywhere throws powerful veins into the encircling slate-band around the borders of the northern district, greatly altering the slate along the line of junction, and disturbing its stratification. The slate is manifestly heaved up by it; and in some vertical sections, as at Tornidneon, and the hill-sides north of Glen Catacol, we have slate above and granite below, with numerous alternations where the two rocks approach. The granite was, therefore, injected in a molten state amid the already formed strata of slate. But, further, it was suggested long ago by Murchison and Sedgwick, in their celebrated paper on Arran (*Geol. Trans.*, vol. iii., second series, 1835), that the bed of limestone on the north front of Maoldon may have once been continuous with the Corrie stratum, and that its actual position is due to an upthrow by a protruding mass of granite advancing from the central mass east of Goatfell. If this be admitted, it would make the intrusion of the granite posterior to the deposition of the carboniferous formations, and so render probable an identity of age between the Ploverfield granite and that of Goatfell. We think, however, that this view is liable to question; we often find exactly similar bands of limestone amid the strata of coal sandstone, without any evidence of former continuity, or such cause of disturbance; and the amount of vertical displacement implied in the supposition could hardly have taken place here without a fracture of the crust, and the appearance of granite on the surface,—so narrow is the band of sedimentary strata superimposed upon the granite. A positive conclusion, then, seems scarcely justifiable from this case. A stronger presumption is derived from the high angle generally assumed by the sandstones, where they approach near the granite, on account of the narrowness of the slate band, and from the degree of metamorphism which in such situations they exhibit; a good example of which occurs at the junction in the burn of the White water, above Corrie, where a gradual passage takes place from slate to sandstone, clearly the effect of metamorphism, by the heat to which both were subjected. The facts all tend to shew the posteriority of the granite outburst to the deposit of the old conglomerate, and that the entire slate stratum on the east or Corrie side was in a plastic state, under the influence of the intense heat which fused the granite.

Viewing all these facts in connection with the general conformability of the carboniferous strata to the old red sandstone, and the gradual transition from the one series to the other, observed in several places, there seems a great probability that the injection of the granite, in a melted state, among the strata took place after the deposit of the carboniferous formations; and that therefore the granites of the three disconnected tracts may be all of one age, or belong to the same period of disturbance. But as the granite of the nucleus is nowhere seen to alter the carboniferous formations, while it certainly does, as above stated, alter the old red sandstone, it is quite possible that these carboniferous strata may have been deposited upon the old sandstone during a period subsequent to the irruption of the granite. But this irruption took place in hypogene depths, not only prior to the elevation of the island above the waters of the primeval ocean, but while the granite was yet enveloped by the mantling slate rocks, and perhaps also by the later formations. It is obvious, as already pointed out in Art. 9, that these secondary strata have not derived the detrital materials of which they are made up from the disintegration of the granite; this rock was yet protected from disintegration by its mantle of slate; and the old red derived its materials from other, and some of them remote sources. An extensive disintegration and denudation may even have gone on for a long period, ere yet the strata were injected by the molten granite; for in several places the conglomerate, partly made up of slate fragments, is altered as well as the underlying slate; and besides, the extreme narrowness of this band on the east renders it very improbable that portions of the injected veins, with adhering slate, should not be found in the conglomerate, if the injection had been prior to the denudation and to the deposit of this latter rock. Such a negative argument, however, has not much value.

The elevation of the central granite mountains to their present height may have been a gradúal process, during the continuance of which, in waters constantly becoming shallower, the strata of slate may have been exposed to further extensive denudation, which, joined to various atmospheric influences, afterwards acting, would give their present form and outline to the jagged ridges of the northern mountains. Long before this elevation took place, the granite, under the pressure of the superincumbent slate, and perhaps of the newer formations also, had acquired its crystalline structure by the slow passage of its heat of fusion into the adjoining strata; and most probably it was quite solidified anteriorily to its elevation, so that it was protruded in a solid form.

14. The agent in this protrusion may have been a newer granite, produced beneath the former. Let fresh accessions of molten matter—the matter of granite—be slowly and constantly transfused from the nether depths, amid the basement portion of the older granite, already cooled and crystalline above, while fused below by contact with the molten mass—this latter will expand, and perhaps laterally extend the former, and raise it in a solid form. Thus a great upward movement might be produced, forming the high mountain nucleus of the north, and at the same time elevating and contorting the strata of slate and sandstone resting on the flanks of the older granite, and in some places perhaps even inverting the dip of the slate, as being subject to a greater strain, and more likely to yield *en masse*, without disruption; while the sandstones of the southern plateau, remote from the focus of intensity in the upheaving force, would be elevated from below in more horizontal strata. The newer granite below might likewise impenetrate the older, and so make its appearance in lower situations, when the land was finally raised, and had assumed, in virtue of denudations

effected during the process and by causes afterwards acting, somewhat of the aspect which it now retains. Formed under such conditions, this later granite might be expected to differ in structure, if not also in composition, from the older granite invaded and displaced by it. Such differences we know do actually exist between granites in the Alps, Andes, and other localities, which can be clearly proved to be of different ages. Now, we have two such granites within the area of the mountain nucleus, the coarse and the fine-grained (Art. 6). With this latter, Mr Ramsay's Ploverfield and our Craig-Dhu granite exactly agree (Art. 11, 12). Is it not, then, probable that these three belong to one period of disturbance—that they were simultaneously injected amid the rocks which now enclose them, at a period subsequent to the deposit of the carboniferous strata? This conclusion is rendered highly probable by the character of the contact between the fine granite and the slate, on the south boundary near Loch Ghnuis, to which reference has been made (Art. 6). The two rocks adhere firmly, the slate is altered in the usual way, and invaded by ramifying veins, in the same manner as the coarse variety invades the slate in other places. At several yards back from the junction, a few masses, or large blocks of the coarse-grained variety are seen, and in two or three cases were observed to be penetrated by veins of the fine kind; in others the two kinds were irregularly associated in one block, in a way suggesting either injection of one into the other, or a simultaneous crystallization under varying conditions, the true relations being masked by the extensive decomposition which has affected both. The junction here was traced for upwards of a mile, and the coarse and fine kind found to be irregularly associated throughout, the fine predominating, and the coarse often occurring in such positions that it was difficult to determine whether it was *in situ*, or transported. The contortions of the slate are very striking along the boundary. The appearances here, in connexion with the other relations already noticed, certainly favour the idea of a late intrusion of the fine granite. Some importance, however, must be attached to the occurrence of the rounded pieces of fine-grained granite enclosed in the conglomerate, close to the Craig-Dhu granite (Art. 12). If our explanation of their occurrence there be deemed unsatisfactory, then will not the above conclusion hold; and it must be admitted that, before the irruption of the Craig-Dhu granite, the fine-grained variety of the interior must have been elevated, stript of its slate covering, and exposed to degradation.

15. What, then, it may be asked, is the conclusion which we favour, and to be finally drawn from these various and somewhat conflicting statements? The discussion may seem tedious and unimportant to many; yet we hope it will not be without its use to the student and future inquirer; and as several of the facts are new, it may have some value in the eyes of the many geologists in this and other countries who have either written upon the subject, or take a lively interest in the physical history of this extraordinary island. The question of relative age is, we hope, much narrowed by these statements, but for the present must remain unsettled. The various possible conclusions may be set forth, by way of recapitulation, as follows:—

1. The oldest rock in the island is the slate.

2. The old red sandstone, and the carboniferous sandstones with their intercalated limestones and coal strata, were formed before the granite was exposed to disintegration, the only fragments of this rock yet found being those in close proximity to the Craig-Dhu mass, of which they are probably injected pieces, and not derived from the disintegration of a granite already exposed.

3. The injection of the granite of the nucleus, whether coarse or fine, in a molten state, amid the slate strata, was certainly posterior to the deposit of the old red sandstone, and may have been posterior also to that of the carboniferous strata.

4. If the granite of the nucleus be thus of later age than the carboniferous strata, then may all the granites be of one age, if such differences of mineral structure, or aggre-

gation of parts, can be admitted to exist in contemporaneous granites.

5. But as the coarse-grained granite cannot with certainty be pronounced newer than the carboniferous strata, while the Craig-Dhu and Ploverfield granites undoubtedly are so, then we may have two ages for these outbursts—one for that of the coarse granite, and another for that of these latter.

6. The constant character of the fine-grained granite of the interior, through a considerable area, its subordinate position to the coarse, and its impenetration of the latter, and of the slate in veins, point to a later origin than that of the latter; while its almost perfect identity in structure and arrangement of parts with the other two granites, render very probable the cotemporaneity of these three, and their posteriority to the coarse-grained variety. This conclusion seems now warranted by the facts; still it would be satisfactory to trace out the junction of the two granites of the nucleus, and ascertain the existence of similar veins. Already, we think, sufficient evidence has been set forth (Art. 6), to shew that this fine-grained granite is not the newest rock in the island, but is older than the traps and porphyries.—To a short history of these we now proceed.

The Trappean Rocks

16. Arran is extremely rich in rocks of this class; most of the known species occur, and also those numerous varieties by which these graduate into one another. They form great overlying masses, capping the sandstone of the southern plateau, and rising into the highest hills of this division of the island. They are interposed amid the sedimentary deposits, in huge sheets or beds conformable to the stratification, and cut through all the rocks alike, from the lowest to the highest, in vertical or slightly inclined dikes, which range continuously across great horizontal distances, in one uniform direction. These dikes are never observed to wedge out downwards; and no doubt they descend to great depths below the surface, where

sheets of molten matter still exist, concentric with the crust —the common source whence they all proceeded, and whose vents or outlets these dikes once formed, in past stages of the earth's history, when the various rocky materials were elaborating. The pressure of an ocean of great depth, or that of other strata, amid which they were poured out, gave these various igneous products that density and compactness, which constitute almost the sole differences between them and the modern products of fire thrown out under the pressure of the atmosphere only. They differ little in

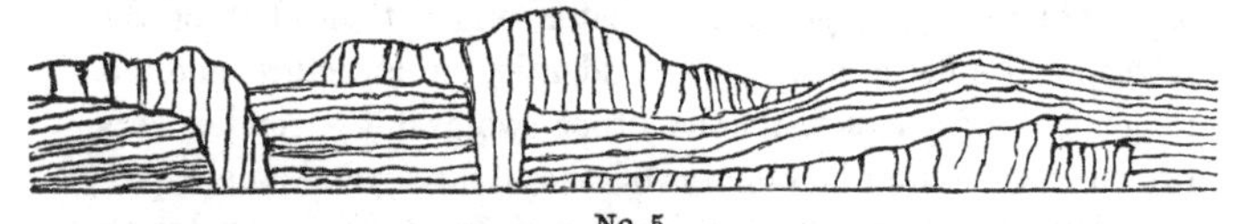

No. 5.
Relation of the Igneous and Sedimentary Rocks.

their chemical composition—mainly in that arrangement of component parts which would be given by different rates of cooling from a state of fusion.

The trap rocks of Arran may be arranged in three classes, according to their composition,—

The Felspathic, comprising porphyry, claystone, compact felspar, and pitchstone.

The Hornblendic, as basalt, greenstone, clinkstone, and amygdaloid.

The Hornblendo-felspathic, as syenite and trap-porphyry.

These are all intimately connected, one species often passing into another by regular gradations, and they are all found in the same relative positions with respect to the sedimentary strata. Among themselves they do not preserve any order of succession, nor do they occupy separate areas, so that their continuity cannot be reckoned on through a considerable space. Nor can they be indicated on a map by distinct colours, so intimately are they blended with one another. Basalt, greenstone, and trap-porphyry are by far the most abundant, as well in overlying masses as in dikes and interposed beds. Porphyry is next in abundance, and occurs in all these positions; pitchstone alone has not overflowed the surface, and occurs only in dikes and beds. The overlying masses are limited to the southern section of the island, while dikes occur everywhere, not however with the same frequency in all parts. On this subject Professor Phillips was, we believe, the first to offer a good generalization—"Dikes are most abundant at some distance from the granitic centre. At Corriegills, at Lamlash and Tormore, they are exceedingly abundant in the red sandstone, while in the north-eastern face of the island, where that rock is nearer to the granite, fewer dikes appear; and about Loch Ranza the slate is still less divided by them. Perhaps we may venture to add another generalization,—viz., that these dikes are most abundant beyond the line of violent flexure of the strata from their horizontal position. After measuring with care the directions and breadths, and noting the characters of forty-four dikes, chiefly of greenstone, between Brodick and Lamlash, and also those at Tormore, it did not appear to us that any other dependence of the direction of these dikes upon the local centre of the granitic eruption could be traced."—(*Man. of Geol.*, 1855, p. 505).

Rocks of the felspathic types, which are most closely allied to granite, seem to have no more intimate relation to the granitic centre than have those of the hornblendic. The largest body of porphyry on the island is that of Leac-a-breac, on the south-west; the next in extent is that of Dunfion, over the Corriegills shore. A similar rock occupies a small space on the Windmill Hill, over Glen Cloy, in close connection with the Ploverfield granite. A different variety forms the bold precipices of Drumadoon, on the west, and the principal mass of Benanhead, on the south. The largest body of claystone forms the middle and upper portions of Holy Isle, and has a thickness of nearly 900 feet; extensive beds and dikes of the same substance are met with in Lamlash river and at Blackwater-foot; lesser veins and beds in many other places. All of these cut through or overlie the carboniferous formations of the southern section of the island; the northern section, the region of granite, slate, and old red sandstone, is almost devoid of these felspathic rocks; a few dikes only are met with; almost all those in this tract being of hornblendic rocks. The pitchstones also, exclusive of those in the granite, are almost all met with in the neighbourhood of Brodick, and towards Mauchrie water on the opposite shore. These various felspathic rocks thus seem to correspond pretty nearly on opposite sides of the island, and to have no relation, in their position, to the granitic nucleus. Neither does there seem to us any good foundation for a generalization put forward by some writers on Arran, that rocks of this type are more abundant on the western than on the eastern side of the island.

Overlying rocks purely hornblendic, as basalt and greenstone, occur chiefly over the central and south-eastern portions of the southern plateau, south of the parallel of Lamlash. North of this line too great an extension has been hitherto given to these rocks; they merely cap the sandstone in isolated knolls or narrow bands of inconsiderable thickness. The details regarding these, as well as the felspathic rocks, will be seen upon the map, and will be more fully noticed in the several "Excursions" which follow. Under the same head we shall notice the changes made by the dikes on the adjoining rocks.

In Arran, as elsewhere, almost all the dikes are simple—that is, composed of one kind of rock; while a few—of which the most remarkable are those of Tormore—are composed of parallel bands of different substances. By far the greater number, in every part, consist of common trap, that is, some variety of greenstone or basalt.

17. M. Necker, who visited Arran in 1839, carefully measured the direction by the compass of a great number of dikes seen along the eastern shore and in the interior, and laying these directions down upon a map, he identified a great many of them at remote points. Such continuity has

been often made out through distances greater than the whole length of Arran; but we know of few tracts pervaded by dikes, where the same caution is required in drawing such conclusions, on account of the great number of dikes, their varying directions, and the undulations to which the same dike is subject. The prevailing direction is towards the north-west and north-east quarters, and nearly within the limits between which the magnetic needle is known to vary. Yet are there many which range without these limits, and not a few have a direction nearly due east and west. Most of the dikes are vertical; those which are inclined to the horizon seldom pass an angle of inclination of 20°. M. Necker estimates the number of dikes between Loch Ranza and King's-cross point at 200; the number to the south of this, on the east coast, at 144, making a total of 344. The remaining portion of this estimate we give in his own words:—

"Mais l'évaluation précèdente ne comprend que les dykes de la surface d'une moitié environ de l'île; tout l'intérieur de la partie meridionale n' y est pas compris, non plus que la côte N.O., ni le groupe granitique de Ben Vearan entre cette côte et la rivière Irsa; et quoiqu' il soit connu que l'intérieur des terres renferme toujours moins de dykes que les côtes, et que la côte N.O. est en général très dépourvue de dykes, quoiqu' enfin cette moitié de l'île soit bien plus petite que cette que j'ai parcourue; omettant ces circonstances je porterai pour elle un nombre égal à la première, soit 344, formant un total de 688, ou, en nombre rond, de 700 dykes de trap dans la totalité de l'île d'Arran. Doublant même encore ce nombre si l'on vouloit, pour y comprendre tous les dykes cachés par les bruyères vastes et étendues dans l'intérieur, par les grèves de sable sur les rivages, ou placés dans des recoins inaccéssibles des montagnes, on n'arriveroit pas encore au nombre de 1,500; et pourtant en parlant de telle ou telle côte, de telle localité d'Arran, il est souvent échappé à ceux des géologues qui ont decrit Arran, à moi-même peut-être tout le premier, de dire qu' on y voyoit des *innombrables* dykes de trap. Or, je crois avoir maintenant montré que loin de ne pouvoir être comptés, on peut à présent concevoir l'esperance de voir chacun des dykes de cette île individuellement étudié, numéroté, décrit et enregistré dans un catalogue descriptif et raisonné, analogue à celui que j'ai aujourd' hui l'honneur de mettre sous les yeux de la Société Royale."*

The hope expressed in the last sentence will soon be to some extent realized. All the dikes are to be laid down in their true bearings upon the large Ordnance Map now in process of construction.

Mr James Napier, of Glasgow, has published a short paper on the dikes between the bays of Brodick and Lamlash in the *Edinburgh New Philosophical Journal*, New Series, vol. ii. No. 1, July, 1855, accompanied by a map, on which the dikes are laid down. He reckons altogether fifty-four dikes as visible along the shore, but considers that not a few may have escaped his notice. "Struck," says Mr Napier, "by the large number of trap dikes cutting through the sandstone, in a direction at right angles to the sea-line, it occurred to me that if such dikes continued round the coast to Lamlash, and still at right angles to the sea-line, they must in all probability have proceeded from a common centre, lying somewhere between the two bays." To test this idea by observation, he measured and marked down the position of every dike, and the result confirmed his "anticipation, that they proceeded from one, or possibly from two centres." A similar idea would be very likely to occur on examination of M. Necker's map, which certainly Mr Napier had not seen, else he would have mentioned it. The notion of radiation from a common centre we do not, however, find alluded to in M. Necker's paper. Mr Napier seems disposed to assign two centres—one for the felspathic dikes, and another for the hornblendic—both lying inland towards the Lamlash road. Prolonging the directions of the two principal felspathic dikes on the shore,

* Documents sur les Dykes de Trap d'une Partie de l'île d'Arran. *Transactions of the Royal Society, Edinburgh*, vol. xiv., Part 2, 1840, p. 684. The paper is a model of patient and generally accurate research.

he finds that they would meet near the claystone quarry on the Lamlash road, about a mile from Springbank; and here he would place the felspathic centre; the hornblendic he does not so definitely fix.

Now, analogies in support of this view can certainly be drawn from districts of recent volcanic action, where fissures radiating from a vent, or focus of disturbance, are seen to be filled with basaltic lava and other igneous matters; and the same may doubtless have occurred in the case of the plutonic rocks; but the evidence for it in Arran we cannot consider sufficient. There are many exceptions to the rectilineal course of dikes here as in other places; some of the dikes converge towards the Corriegills shore, and the largest runs a long way parallel to it, while one at least re-appears far inland beyond the place of the supposed focus. Besides, so far from "the whole of the hills between Brodick and Lamlash being composed of trap," this rock is, in point of fact, confined to a narrow and thin capping along the highest ridge between the two bays; and Mr Napier has overlooked the great outburst of porphyry at Dunfion, which has a manifest relation to the felspathic dikes on the shore, as well as the numerous masses of claystone intercalated amongst the sandstone strata along the northern slopes west of Corriegills.

On these grounds we cannot admit that this speculation has much value; the apparent radiation arises from the circumstance already mentioned, that the vast majority of the dikes range between the points of extreme magnetic declination east and west. Some other suggestions, however, of this paper have great value; those, namely, of a chemical nature, referring to the different degrees of fusibility and rates of cooling among trap rocks, which, if experimentally established, would elucidate many points still obscure in the natural history of the ancient products of plutonic fires. In a later paper, describing the Whiting bay dikes, Mr Napier forms a much higher estimate than M. Necker of the total number on the east coast (*Trans. Phil. Soc. Glas.*, Vol. IV., p. 321).

Glacial Action

18. The various accounts which we possess of the geology of Arran, and the separate memoirs upon it, were written before glacial action had been recognized in the production of superficial phenomena, and we are not aware that any geologist has since turned his attention to the subject. Those phenomena, indeed, had been observed of which the ice theory is now considered to offer the most satisfactory explanation; but the island has not hitherto been examined with the view to discover the direct evidence of the action of ice, such as striated and polished rocks, and "roches moutonnées." Visiting Arran in 1855, after having spent two months of the previous summer in a district abounding in unequivocal evidences of such action, I felt the greater confidence in undertaking this task. The result may now be briefly stated; details being reserved for the notices regarding particular tracts, which will be given farther on.

The remarkable peak of Cior-Mhor, whose altitude is about 2,700 feet, has been already (Art. 2) pointed out as the geographical centre of the northern group of mountains. From its base the four great valleys of the island—Sannox, Rosa, Iorsa, and Eais-na-Vearraid—radiate in all directions, their extremities opening on the seaward belt of low land. If glaciers ever existed in Arran, under the subarctic climate to which Scotland was once subjected, these central heights must have been the seat of the snow-fields which fed them, and the radiating valleys the channels down which the viscous mass of glacier ice must have pushed forward to debouch upon the low ground, and melt under a higher temperature. On the sides, then, and towards the openings of these valleys, we should expect the effects of glacial action to be most distinctly traceable in the striation and polishing of the subjacent rocks and transported masses, and in the formation of lateral and terminal moraines. Many broad surfaces of the natural rock are exposed both on the sides and in the bottoms of these valleys, favourably placed for receiving such impressions under the grinding action of a descending mass; yet have we failed in detecting more than a few cases of striation and polishing, or of that "moutonnée" character of surface; which is referable to the action of moving ice. A granite surface is, however, very unfavourable for the preservation of such markings, especially the Arran granite, which is generally of such structure as to be subject to rapid disintegration. The slate is better fitted to retain impressions of this kind, its toughness and fine-grained structure rendering it less liable to decomposition; but it is seldom exposed in favourable situations, and is rarely found striated. Granite bosses in the glens, and on many of the lower ridges, have that peculiar rounded character, due to the action of ice, to which the term "roches moutonnées" has been applied; but perhaps none of the cases can be decidedly referred to glacial action, on account of the peculiar spherical structure so often assumed here by granite on the large scale. On the slate ridges, however, beyond the granite border, some well-marked cases do occur, as on the plateau to the south-west of Goatfell.

But though these more direct evidences are so rarely met with here, there are others scarcely less satisfactory, and of more frequent occurrence. These are the terraces and mounds of transported materials on the sides and at the openings of the glens, and the dispersed blocks in every part of the island.

The terraces and mounds consist of earth and rounded masses of rock of each particular glen, irregularly mixed without reference to weight, and in such situations, that they could not have been brought together by existing river action, being much above the level of the streams which now traverse the valleys. They are most probably referable to glacier moraines of two classes—the lateral and terminal—formed by masses of rock descending from the highest peaks, and thrown to either side by the movement of the ice, or deposited at the extremity of the glacier when the ice melted. The former have been much modified by torrents entering from the sides, after the glaciers disappeared, and now present but detached mounds. In some glens, however, as Sannox, terraces yet remain complete, but not of great extent. The terminal moraines are better marked, fine examples being visible at the mouths of Glen Iorsa, Glen Catacol, Glen Rosa, and others. The remarkable terraces at the opening of Glen Catacol skirt the valley on the south-west at a height much exceeding any level the stream could now reach by the joint effects of floods and high tides, and indeed surpassing that which it could ever have attained even when the sea covered the present maritime belt or terrace. We are, therefore, inclined to regard these mounds and terraces as terminal moraines, modified in their outlines by floods, tides, and ordinary river action.

Still more remarkable are the lofty terraces at the mouth of the river Iorsa; they are far more striking, indeed, than anything of the kind in Arran. They consist throughout of transported materials, some of the rocky masses being very large; the sides are steep and the summits usually flat; and the height of the highest is sixty or seventy feet above the river, and at least thirty above the ancient sea level just alluded to. Speaking of these (Iorsa, and other such mounds) in reference to river action, MacCulloch remarks:—"The origin of such alluvia is very obscure—a few may have been deposited in particular situations by the same waters which are now removing what they formerly laid down; while in other cases it is impossible to assign any mode of action by which this double and opposite effect could have taken place from one agent. . . . The quantity and quality of the materials, their extremely rounded forms, the nature and permanence of the hills above, and the want of a regular gradation of size in the stones from the bottom upwards, seem to show that other causes [than river action] of a transient, and probably of a diluvian nature, have in distant times generated these deposits, which have been subsequently acted on by the stream concentrated on the bottom of the glen by the form of the ground" (*Western Isles*, ii., 335, 1819).

The difficulties of these cases had thus presented themselves to the mind of this distinguished geologist, and he offers for their solution a *vera causa*, one certainly capable of producing such appearances. At that time glacial action had not been recognised in these countries; and he offers the best explanation that could then have been given. The action of ice is, however, more simple, rational, and consistent, not only with these appearances, but with others to be presently mentioned.

Similar mounds occur, but not in the terraced form, at the openings of Glen Sannox and Glen Rosa, much elevated above the river beds. Some way up the latter, also, there are remarkable mounds, in a situation where, from a great bend in the glen, we should expect a moraine to be thrown down. Many other examples might be given; but it is unnecessary to refer in greater detail to phenomena of this class. The paucity of decided cases of striation—such as are so frequently and distinctly marked in the valleys of the lake district of Cumberland, or as are seen on many parts of the Clyde shores, and in the southern Highlands—goes so far against the glacial theory as applicable to Arran, and in favour of the idea that the mounds and terraces in question were formed when currents swept these glens, during the gradual elevation of the land. That such elevation may have been a long-continued process we have already seen reason to suppose (Art. 13, 14); and the effect must have been a general disturbance in the sea bed, which, joined to the action of tides produced then as now, could not fail to give rise to currents of considerable force. Where these met the sea, towards the mouths of the glens, banks and terraces may have been thrown up; or a sudden elevation of the land of a cataclysmal character may have given origin to long-continued currents of sufficient force to transport large blocks, and to throw down a promiscuous deposit, such as we find in the mounds and terraces of the mountain glens. Such sudden elevations of a range or group of mountains are still regarded by many geologists as the true explanation of the "diluvial phenomena;" and it was such probably that Dr MacCulloch had in view when he spoke, in the passage above quoted, "of other causes of a transient, and probably of a diluvian nature," as giving origin to the remarkable accumulations at the mouths of the Catacol and Iorsa.

19. The dispersed blocks present phenomena still more curious and of much more difficult explanation. They are almost exclusively granite, a very few only of slate being found. They are scattered in great numbers over every part of the island, and are often of enormous magnitude. They are most abundant and largest in the vicinity of the granite nucleus, as about Corrie and the shore at Corriegills; and generally less numerous, and of smaller bulk, in the remote southern districts. Occasionally, however, some very large ones occur even there. They are limited to no particular locality, but occur alike in the valleys, on the summits and northern and southern slopes of the hills, in situations to which they must have passed across deep and narrow glens. They are found also isolated on the Holy Isle, which is separated from the mainland by a wide bay, and two deep navigable channels. Blocks of the coarse-grained variety are much more numerous than those of the fine. The latter, indeed, are in a great measure limited to the tracts on which Glen Iorsa, the principal seat of this variety, opens towards the south; and this fact, in connection with the more sparing distribution of the blocks along the northern coast, on which but one glen with a narrow opening debouches, than in other parts of the island, shews that, though the dispersion has been quaquaversal, it has been to a large extent determined by the direction of the valleys. Dr MacCulloch, who has noted the leading facts regarding the dispersion of the granite blocks with great accuracy, though imperfect in many of his details, closes his account with the following observations:—"None of the blocks have the marks of a distant origin; all have the characters of the granites of the adjoining mountains, char-

acters sufficiently distinct from those of almost all the granites of Scotland. . . . . . . No situation, perhaps, has been pointed out where the origin of the travelled blocks is more obvious, or their new position more difficult to comprehend, without assuming considerable revolutions of the surface of the land over which they have passed. . . . . . The compact and solitary position of the fixed mass of granite, the identity of the materials of this mass with that of the travelled stones, the gradual diminution of these as they recede from the parent rock, and the insulated position of the whole, render their origin indubitable, and present to the geologist a spot, on the changes of which he may speculate, with the certainty that he has before him a set of incontrovertible data from which to reason."—(*Ut sup.*, p. 341.)

This passage places in a clear light the conditions of the problem, and the difficulties attending it. The author does not, however, propose a solution of the difficulties, nor does he enter into any theoretical discussion. His account of the travelled blocks is the only one which we have seen; no other writer on Arran, that we know of, has turned his attention to the subject. It is hoped, therefore, that the notices now given will be the more acceptable.

In Arran, as generally in other districts, the boulders belong to a particular period. The entire system of rocky strata had been formed, and the existing inequalities of the surface established; but in all probability the last upward movement of the land, to which we have already often referred, had not taken place. The relative age, in fact, seems to have coincided with that of the boulder-clay of Scotland, or with the newer pleiocene era. Then, as regards the forces concerned, we know only two natural agents capable of producing the effects. These are currents of water and moving masses of ice. Now, the former are totally inadequate to carry forward masses of the enormous magnitude found here, or even to transport the lesser blocks over all the obstacles which they have surmounted, in their outward course from the parent rock. Besides, they are often found "perched" in situations where it is extremely improbable that currents could have left them, and also crowded together in groups in places quite open, and removed from the influence of eddies. It is true, indeed, that the origin of such currents can be readily accounted for, by movements which we know to have taken place—the elevation, namely, of the mountain nucleus from beneath the sea. We have only to suppose that it was sudden and of considerable amount, and we have at once generated a series of mighty pulses, which would carry the disturbed waters, with their load of torn off materials, along the surface of the lower lands still submerged. Rocky materials may thus have been swept away, and re-arranged in new situations, valleys scooped out, and extensive denudations effected. But the forces thus brought into play cannot have been adequate to bear along the enormous masses, now far separated from the parent rock; and therefore we do not hesitate, on this and the other grounds above stated, to conclude that moving masses of ice were the transporting agents. In the passage of glacier ice adown the valleys, and the buoyancy of floating bergs, forces of sufficient energy would be lodged to carry the largest masses; and this agency we know is adequate to produce the various phenomena of transport, grouping, and "perched blocks."

20. We thus seem shut up to the conclusion, that the agent in this transport was ice in motion. Now, this agency may have been brought into play in two ways. The northern mountains may have formed a mass of ice-covered land, with glaciers descending to the bold shores of the sea of that period, while the southern plateau may have been under water. From the extremities of the glaciers, masses of ice, of which some must have been considerable bergs, and the sea therefore not very shallow, would float away, carrying the granite blocks which had fallen on the ice towards the heads of the valleys, and been borne along on the glacier. Stranding or melting, these floating masses would throw down their load of blocks, and thus the shores of the island and the surface of the south-

ern plateau may have become encumbered with vast multitudes of granite blocks, chiefly of that coarse-grained variety which constitutes all the highest mountains.—But in another way ice may have been the agent of the transport in question. The whole island, elevated in both divisions high above the waters, may have been wrapped in sheets of ice, across which the granite blocks, as they dropped from the high peaks and precipices on which the snow could not rest, or were torn off by the pressure of the ice from the sides of the glens, would be carried onwards in all directions by the slowly descending viscous mass. Existing glaciers do not require very steep slopes for accomplishing the transport of large masses. A moderate inclination is sufficient, and we need not therefore suppose that there existed any very great difference in the relative levels of the surface in Arran from that which now obtains. All the largest blocks are found at the lowest levels, and comparatively near the granite nucleus—the lesser may have found their way across the surface of the southern plateau, the valleys being all filled with ice, borne along with the slowly advancing mass.—Our views formerly inclined to the iceberg theory and submarine deposit of the blocks, but a recent careful examination of the lake district of England, and a comparison of the markings with those of Switzerland, have led us to the conclusion that the agency of floating bergs is insufficient to have produced the regularity and persistency which the markings, and other evidences of ice action, now present; and that an icy envelope in a state of constant advance will alone explain them. This same force is sufficient greatly to alter the forms and dimensions of valleys, if not actually to scoop them out, independently of any pre-existing dislocation, or fault, such as is considered to have determined the first action of water; while it is generally allowed that on this view we best account for the great detrital accumulations at the mouths and along the sides of the mountain glens. This supposed state of the surface of the land implies the existence of temperatures in the atmosphere and in the waters of the adjoining sea, such as would favour the development of an arctic fauna in the waters —a condition of things, indeed, like that now prevailing in parts of Greenland, where the ice, which covers the land, sends down glaciers to the sea level; and underneath the rim of ice which fringes the coast, a peculiar group of testacea flourish, very different from those of the British seas. Now, such an assemblage of arctic species of testacea actually exists in many parts of the basin of the Clyde, and along the shores of its friths, and has very recently been found in Arran also, but elevated to various heights, sometimes several hundred feet above the present level of the sea, the species being now extinct in these islands, and only known as denizens of the Greenland and other arctic waters. The conclusion is perfectly legitimate that, coincident with the disappearance of the ice, the land sustained a general elevation, which placed the shelly deposits, elaborated beneath the waters, high above their level, and introduced the temperature and general conditions which, while proving destructive to many species, favoured the immigration and development of that assemblage of species by which the marine fauna is now distinguished. But the present level of the shores was not yet attained, nor the actual coast outlines as yet carved from the rocky border which broke steeply down all round the island. The sea covered the plains of Brodick and Shiskin, and stretched its winding arms far up the solitary glens. During the slow progress of perhaps forty centuries, the streams from the rugged mountain sides and gentler hill-slopes bore down detritus of granite-sand, slate, and quartz pebbles, and spread them out below the waters of the quiet friths. In sheltered places the tides and waves cut a low but well-marked margin along the highest water line; while on the open shores the heavier surge wore deeper, the hill-slopes were cut into a steeper and higher cliff, and hollowed out into caves in all the rocks alike. The testacea and other denizens of the present shores already inhabited the waters of that remote period; but we have no evidence that man had yet appeared. It is most probable, indeed, that the last elevation of the land, to which we have already often alluded, took place before the human period. We are only certain, however, that all the existing levels were established prior to the Roman invasion. Here, as generally in the West of Scotland, this last elevation amounted to about forty feet, and gave to Arran its present maritime border, and the inland cliff which forms a singularly picturesque feature in its coast scenery.

Shell Beds

21. The occurrence of recent shells of arctic species in the deposit called the "boulder-clay," shows that in Arran, as on the mainland, a climate prevailed favourable to the development of glaciers. This boulder-clay or "till," has long been recognised as existing in Arran; it is seen in fine section in the lower seaward portions of many of the glens, especially in the south and south-west of the island, and is probably spread out, though in thinner layers, over much of the surface between the river channels, which are of course best adapted for the retention and exhibition of detrital matter. The deposit consists of several beds, with distinct and persistent characters, which are often overlooked, so that the whole is regarded as one formation. The lowest portion next the natural rock is an unstratified clay of a red or chocolate brown colour, in which large boulders, both angular and rounded, and usually striated, are thrown together pell-mell. Over this are beds of laminated clay and unstratified sandy clay, with very few boulders, but occasionally striated pieces of rock. This is the chief repository of the arctic shells; but over it are other beds of a character somewhat different, among which occur shell beds with species such as now inhabit our own shores.

Mr Smith, of Jordanhill, was the first (Mem. Wern. Soc., Jan. 26, 1839,) to distinguish between these shelly deposits—to assign them to two distinct ages—to recognise the arctic character of the lower, and thence to infer that a sub-arctic climate once prevailed in Scotland, connecting with that of the shells other evidences of glacial action. This was an important generalization; it at once introduced order into the heterogeneous mass of beds, ranged conflicting facts and appearances harmoniously under one head, and formed a new era in post-tertiary geology, like that established among the tertiary beds by the generalizations of Lyell.

It is only within the last few months that boreal shells have been discovered amid these superficial beds in Arran. The credit of this interesting discovery is due to the Rev. Robert Boog Watson, B.A., F.R.S.E., of Edinburgh, who has kindly furnished the annexed list of the shells. A detailed account of the beds and their fossil contents was read by Mr Watson to the Royal Society of Edinburgh in April last, and published in abstract, but without the list, in their "Proceedings," Vol. V., No. 63. We have since visited the chief locality, which is on Torlin water, about one mile inland, and twenty or thirty feet above the stream; and found seven of the species, in an hour's digging, besides two species of foramenifera. The following is a list of the shells:—

Balanus crenatus.
Purpura Norvegica.
Tellina Baltica, a brackish-water variety of solidula.
Cyprina Islandica.
Astarte elliptica.
——— arctica.
——— compressa.
——— striata, a very large variety of compressa.
Cryptodon Sarsii.
Modiola modiolus.
Leda pygmæa.
—— pernula.
Pecten Islandicus.
—— opercularis.
Littorina littorea.
Turritella communis.
Natica —— ?

The perfect state of preservation of most of the shells indicates a quiet deposit on the sea bottom; the elevation of the bed above the present sea-level expresses the amount of former depression below that level; it expresses, indeed, a much greater depression, since many of the shells are exclusively deep-water species: the Cyprina Islandica, for example, is not met with in our seas higher than at a depth of ten fathoms. Further, it follows from what has been stated in regard to the relative position of the beds, that if we can trace an undoubted boulder-clay deposit up the glens to the height of 1000 or 1500 feet, even if we find no shells, the elevation unquestionably indicates this amount *at least* of former depression.

22. The arctic, or sub-arctic, climate which prevailed in the circumstances we are considering, would cover the hill-tops with perpetual snow; glaciers would fill the heads of the valleys, then greatly reduced in length by the submergence of the lower lands, and the marine fauna of the period would flourish under and outside the rim of ice which girt the island; while bergs and floes would float away into the open water, bearing up buoyant for a time, and then throwing down their load of blocks to encumber the southern plateau, as already pointed out (Art. 20). The disturbances attendant on the re-elevation of the land could not fail to produce a certain re-arrangement of the surface beds, and even occasional intermixture, and the heaping up of detritus in peculiar forms, such as those already noticed in the case of Glen Iorsa. The boulder-clay by its position would, in many cases, be protected from such action, while in some situations more exposed its upper surface would be subjected to great erosion or removal, and hence, no doubt, in part the varying depth and undulating surface of this deposit. Its original formation was most probably contemporaneous with the incipient depression of the land, and due to the joint action of ice and water; while its internal structure, huge boulders, unstratified character, and the almost total absence of shells, indicate a period of violent disturbance, a hurried pell-mell admixture and deposit of earth and rocky materials. The angular form of many of the blocks, the perfect and very

general striation of the angular and rounded alike, so that in many places *non-striation* is the exception, clearly shew that, by whatever agency this singular deposit was formed, and the striation effected, there cannot have been afterwards a lengthened transport by rivers, as this is known in the case of the Alpine streams to produce a speedy obliteration of such markings.

Further details regarding this curious deposit will be given in some of our Excursions; and the subject of the shell beds, and the order of superposition of the clays, will be treated at greater length in our account of the basin of the Clyde. Shells from the later deposits, the upper portions of these beds, and from the surface of the marine terrace and caves of the old sea cliff, have been found in a few places. These all belong to species now inhabiting our seas; they indicate that our climate had changed its sub-arctic character, and become similar to that which now prevails.

EXCURSIONS IN ARRAN

EXCURSION I.

To the Summit of Cior-Mhor.

23. This shall be our first Excursion. It will reveal to us the structure of the granite nucleus, and the relations of the mountain groups. It will show us the basset edges of all the strata as we pass in succession across them to the central granite. We shall see some curious dikes, explore a lonely corrie, climb a high pass, and thread a difficult pathway by the edges of the highest cols. The morning clouds have melted from the peaks; there is neither bank of mist nor cirrous haze to veil the far-off horizon; the day will be calm and bright, and the view from the summit glorious. Cior-Mhor (Kior-Vawr) is that far-off peak, with sharp point and rugged shoulders, rising behind the plateau on the south-west of Goatfell. We shall reach it by a walk longer and more difficult than the ascent of Goatfell, and the more exciting that it is somewhat dangerous. Should a mist happen to surprise us upon certain portions of the route, our situation might be very critical. How promising soever the day may be, let the climber never enter these mountains without a pocket compass. The mists come down so sudden and so thick in this changeful climate that without it he will be bewildered and lost amid the high cols and peaks, and huge slippery granite sheets. With ordinary caution, and the use of a correct compass, he has nothing to fear. True, he will find neither house nor herd's shieling within the mountain circuit; but directing his steps

by the compass, he cannot fail, in a walk of two or three hours at the utmost, to reach the inhabited border, where a frank welcome will meet him at every cottage door. But let him bear in mind, that the variation of the compass in Arran is now about $26\frac{1}{2}°$ west of true north, with a yearly decrease of $4'$ or $5'$. To the geologist we would recommend to carry, besides his hammer, indispensable at every step, a good clinometer with which to note the varying dips and inclinations of the slate and sandstone, and the relations of these to the granite centre and the numerous dikes. To the botanist, besides his vasculum, a field book will be very useful, for immediate pressure of the delicate mountain plants. On this and most of our Excursions our departure will be taken from the shores of Brodick Bay, the unrivalled grandeur and beauty of which attract the greatest number of summer visitants. For a few days' sojourn new inducements are now afforded by a well-managed and spacious hotel on a beautiful site at Invercloy, on the south side of the bay. Here, and at Corrie and Lamlash, steamers call several times daily during the summer months.

24. The rock on the shore at Invercloy is a conglomerate of the age of the coal—a member, in fact, of the coal formation. Murchison, Sedgwick, and Ramsay have classed it as a lower member of the New Red; but, as already stated (Art. 6), on evidence which we think inconclusive. The inland cliff marking the old coast line is well seen on the Invercloy shore, and extends far up both sides of Glen Cloy. The lower part of the glen, much of the plain of Brodick, and the marshy grounds at the head of the bay, are but an expansion of the terrace which formed the sea bottom, when the tides and waves were carving out the cliff. The alluvium and rolled stones form, however, but a thin covering to the subjacent sandstone, which appears in the river bed a little way up the glen. It appears also in the bed of the Rosa burn, but nowhere in the Brodick plain. This plain, indeed, is but an old expansion of the beach. The bay formerly had its termination some way up Glen Rosa. The mound in front of the glen, on which a farm-

house lately stood, is but 40 feet above high water; the mouth of Glen Rosa is much less; in fact, it is quite within the limits, to a considerable distance up, of the old sea terrace. Brodick plain shows beneath the soil a continuous covering of rolled gravel cemented by iron, and impervious to water. To drain it in any part, it is only necessary to pierce this covering, when the water at once disappears. On the north-west of Glen Cloy the sandstone is quarried upon the line of cliff, and forms a tolerable building stone. The dip is nearly south, at about 25°. By thé side of Brodick wood, adjoining the handsome new school-house erected by the late Duke of Hamilton, a vein, or perhaps bed, of pitchstone occurs in the sandstone. A portion only of the front appears by the side of a lane, showing a prismatic structure in the rock, and an underlie towards the west. In large blocks lying loose upon the surface, a similar structure is seen. The colour is bottle-green, and specks of red felspar disseminated in the base give the rock a porphyritic texture, approaching that of pitchstone porphyry. The direction is nearly south-east and north-west. It is probably a continuation of one of the many beds or veins of the Corriegills shore, which we shall notice in another Excursion. A breadth or width of fully thirty yards is exposed, but it is very difficult to say whether it is a bed or a vein—it is most probably a bed, as sandstone is seen on the line of its direction, between it and the wall of the wood, and it cannot be found anywhere in the wood, though if a vein it ought to show again, as trap dikes do. No contact with the adjoining rock is exposed. A dike of disintegrating ironshot greenstone, forming a bank behind the school house, has an angular course with respect to the vein; but there does not seem to be any connection between them. In the opposite direction the dike intersects the inclined strata of sandstone, but no marked change is produced. It is in excellent taste that the striking geological features of this spot have been left untouched in carrying out the improvements connected with the erection of the school-house.

By the wayside here, where a sweep of the road gradually opens to us one of the finest views in Arran, taking in the first great reach of Glen Rosa, and its magnificent background of mountains, there stands a huge upright stone, marking, perhaps, the spot where a chief was interred, or where a leader fell in the old days of feud and warfare; or mayhap the scene of some decisive battle with the old Norse invaders. Many such stones are found in the island, but their purpose and date of erection are wholly matters of conjecture. There are several in this immediate neighbourhood; on the high ground south of Invercloy, and on the plain of Glen Shant, between the mouth of Glen Rosa and the site of the old village of Brodick. A complete circle of such stones formerly existed at the mouth of Glen Shirag; it is briefly noticed in Mr Headrick's book; but not a vestige now remains; in 1813 the stones were broken up and removed, to make way for the operations of the plough. In most places where we examined these stones, we found them to be coarse sandstone of the old red formation. As this does not exist in many of the localities where the stones now stand, we must conclude that mechanical appliances of great power were brought to bear in their transport; and therefore it is not wonderful that, in a rude age, their erection was ascribed to the hands of giants. In all ages the illiterate observe facts and phenomena with tolerable accuracy; but their explanations always introduce the marvellous or the supernatural.—The subject will be again referred to in our notice of Tormore.

25. There are two paths to the entrance of Glen Rosa; we take that which passes Brodick Church, and crosses the opening of Glen Shirag. The church stands on a platform, bounded northwards by a low cliff of sandstone, and overlooking one of the most varied and pleasing views in Arran. In this sandstone, underneath the north wall of the church enclosure, there is a bed of carboniferous limestone in a vertical position. It has been largely quarried, and a small portion only is now visible. It contains fossils, of which the most characteristic is the *productus giganteus*, completely identifying this bed with the

limestones of Corrie, the Salt Pans, and Bein Lyster Glen; and enabling us, therefore, to assign the sandstone also, without hesitation, to the age of the coal formation. A little farther on, above a rustic bridge, where the Shirag burn, rushing out from a winding rocky gorge overhung with trees, forms a scene strikingly picturesque, another bed of limestone occurs in the sandstone. Thence to the entrance of Glen Rosa, we pass across the lower beds of the carboniferous formation, which, however, are nowhere seen except in spots in the bed of the Rosa burn. The succession of the strata is shown in the annexed cut.

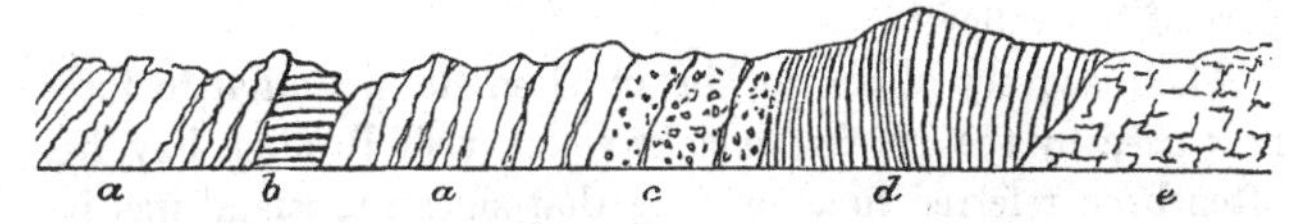

(a a) Sandstone and conglomerate; (b) productus limestone; (c) old red sandstone; (d) schist; (e) granite.

We now reach the outer edge of the band of old red sandstone, which, ranging from the Corrie shore diagonally by the flank of Goatfell, crosses the opening of the glen. It is seen on rocky prominences by the side of the path, but no junction is visible. The beds here exposed consist of a dark-coloured, close-grained sandstone, with specks of mica, bits of quartz, and small clay-galls, and are very characteristic of the upper portions of the formation. The lower portions are seen farther up the glen, but at some height on either side, in the wood and moor on the left, and the hill-side on the right. They are very coarsely conglomerate. Specimens of both varieties may be seen in the stone fence by the side of the wood.

The alluvial mound at the entrance of the glen has been noticed already, as most probably the terminal moraine of a glacier which once filled the valley. It is precisely in the position where such a moraine would have been thrown down, and consists of such materials as the ice would have borne forward; and its height places it far above any existing river action. It appears to have extended, at some former period, entirely across the opening of the valley, backing against the hill-side on the south, as it now does in the opposite direction, and forming the barrier which confined a lake, occupying at that period the lower portion of the glen. The bursting of such barriers, and their subsequent modification by floods in the river, are common phenomena in mountainous districts. Traces of a lateral moraine are seen on both sides; and farther up the glen, where it turns northwards, two other mounds, rising high above the stream, are in the position where a terminal moraine would be thrown down, after the ice had retreated from the lower part of the valley.

The discovery of an anchor in Glen Rosa, similar to those now used by the herring smacks which visit Brodick Bay, has often been referred to as proving that since the island was inhabited by a people far advanced in civilization, the sea filled the valley, and afforded a "trustworthy station for ships." That an anchor was really found, brought to the smithy at Invercloy, and worked up into various articles, we think there cannot be a doubt, from the accounts given by several persons still living, who saw it and handled it. All agree, however, in fixing the locality in which it was found at a place where peat is cut, at a considerable height on the southern hill side, and therefore far above the level at which the waters of the sea stood before that last elevation of the land to which we have so often alluded already. If the discovery, then, be admitted as a fact by the archæologist, it is entirely without that geological significancy which attaches to the canoes found in the holms on the banks of the Clyde and within the city of Glasgow, which will be noticed in the sequel. With the archæologist the explanation may be left.

26. The contact of the lower old conglomerate with the clay slate is not seen in Glen Rosa. The latter rock first appears in the bed of the stream, at the sharp turn where it begins to flow eastwards; but the junction must be farther down the burn. The slate rises high into the hills on both sides, form-

ing on the north the principal mass of Glen Shant rock, called also the Pillar, from a large isolated sheet or prism, standing out detached from the front of the bold cliff. The precipice is about 1100 feet above the river, and forms one of the finest features of this noble glen.

We now approach the base of the series, where the central granite rises from beneath the enveloping slate rocks; and here a celebrated junction occurs in the bed of the stream. The hill sides show the contrast of the two rocks from a great distance, in the bare stony character and loose gray blocks on one part, and the grassy or heath-covered slopes, with dark terraced ledges, on the other; and the geologist is therefore prepared to find a junction somewhere here in such a natural section as the river affords. It occurs about two or three dozen yards below the point where the Rosa burn receives from the west side its only tributary—the Garbh-Alt or Rough burn, which drains the whole eastern side of the Ben-Ghnuis range, and comes down into the glen, bounding headlong across the huge granite sheets in a series of striking falls. The junction has not been so well shown these few years as it was formerly, in consequence of the accumulation of loose masses of granite; still it is sufficiently well seen to make the place interesting and instructive in a high degree, and some future floods in the river may again open it better up. The slate is greatly altered for a considerable distance down the stream, and pervaded by small veins and strings of quartz, and granite in which quartz predominates. The colour of the slate is changed, having more of blue than is usual to it; the structure is altered also, the laminæ are contorted, and present thin bands of different colours, chiefly blue and gray; the latter being purely siliceous, that is, flinty slate or quartzite without the colouring matter, iron or manganese, which exists in the former. The slate on this side of the mountains is generally a dark-coloured, coarse, siliceous rock, showing shining crystalline flakes in fresh fractures; in some places assuming that arenaceous, semi-conglomerate structure which used to be designated as greywacke. Both structures are obliterated on approaching the granite; the colour is bluish, or blue and gray in alternate bands, the structure is extremely fine-grained, and the hardness and toughness are both excessive. These changes, the contortion of the laminæ, or the total disappearance of all stratification, coupled with other modifications not seen here, but to be again noticed, clearly indicate that the schist to a considerable distance from the granite was subjected to intense heat, and remained in such a state of at least semi-fusion as to permit, under the action of chemical forces, a new arrangement of parts, and the permeation and interlacing of veins from the molten rock below. The granite veins are less numerous, smaller, and of varieties differing from the ordinary type of this rock more than is usual in most other junctions with the slate; such might perhaps be seen in the interval of several yards here obscured before the granite itself is reached. The great extent of the altered slate seems to indicate that the strata near the junction are of inconsiderable thickness, and that the granite exists beneath at a small depth, as shown in the illustrative cut at *e*. The slate is here traversed by a basaltic dike, still further modifying its altered structure. The dike intersects the bed of the stream at a small angle, but is seen only on the east bank; it ranges about magnetic north and south, and has a breadth of ten or fifteen feet. We shall meet with it farther up the glen on another Excursion. Rising from the bed of the stream in great rhomboidal masses, divided by partings here inclined towards the slate, the granite extends in a slanting direction up the hill towards Goatfell, and on the opposite side by the Garbh-Alt burn, so that the line of junction almost coincides with the southern margin of the stream. The remaining part of our walk is entirely on the granite.

27. Our path now lies up the steep slope forming the western side of the glen, a little to the north of the Garbh-Alt, which we keep on our left hand. Having reached the summit level we are on the southern slope of the high rugged ridge called Ben-Talshan, which forms the western boundary of Glen

Rosa. In the hollow between it and the lower swells of the Ben-Ghnuis range, the Garbh-Alt pursues its rapid course from north to south, along a granite bed, down a pretty fall, and then between perpendicular walls of granite about twelve feet in height, till, escaping from this rocky barrier, it sweeps round the south end of the ridge, and plunges headlong into the depths of Glen Rosa. The gorge has been excavated along the line of a basaltic dike, which occupies the bottom of the stream throughout, and retires from it at the base of the fall. These dikes are prismatic across; and this structure renders their disintegration much more easy than that of the granite. The amount of wearing in this case is measured by the depth of the chasm, and the distance to which the fall has receded. The stream ran at first on the level of the top of the granite walls, and the fall must have been at the southern extremity of the chasm; the recession would cease, or become extremely slow, when the present situation of the fall was reached, as the dike here retires from the stream. The dike is ten to fifteen feet wide, and ranges nearly due north and south. The rocks adhere firmly at the junction, but the alteration on the granite is not remarkable. Similar cases of the excavation of river channels along the line of dikes are frequent in Arran. As a general rule, fractures or faults determine the course of streams in the first instance; along such lines the excavation is much more rapidly effected.

The steep brow on the eastern side of the wide hollow where we now are, exhibits many rounded masses of granite, presenting the "moutonnée" character of surface, as if moulded by the action of ice: but as no striæ were observed, we can hardly ascribe them to the action of glaciers, as the forms may be due to the effects of disintegration on the concentric structure upon the large scale, so often seen in granite. The bed of the stream, as we pass up, is strewed with many loose rounded masses of pitchstone and trap, indicating the existence of dikes or beds of these rocks among the lofty precipices on the west. Mounting this steep brow, to reach the corrie under the north front of Ben-Ghnuis, we meet with a dike of spheroidal trap in the bed of one of the streams; it is about seven feet wide, and ranges 35° W. of N. The rounded masses of granite here may have received their forms from the long-continued action of water trickling over them, and torrents occasionally sweeping along gravel and large stones. Arrived at this corrie, we are in the midst of a scene wild, lonely, and picturesque. The bare and rugged precipices of Ben-Ghnuis rise high into middle air on the south, with their immense sheets and rhombic masses of granite, from six to twelve feet in the side, piled up, block on block, in massive courses, like the huge rough masonry of giants. The topmost row, broken by clefts and deep gashes, due to irregular disintegration, shows grandly in its perfect definition against the clear sky. Along the front, which sweeps round to form one side of the corrie, there stand out here and there enormous pillars of the rock, detached from the cliff behind, resting on a basis which is rapidly giving way under the active agents of waste in this changeful climate, and threatening a speedy descent into the valley of the Garbh-Alt. The rugged outlines of the Goatfell group bound our view on the east, the distant landscape being shut out by the intervening ridges; and the eye from this point does not take in a single human dwelling, or other sign of the abode of man. No sound reaches the ear but that of the crystal rills trickling from the clefts of the granite, the hum of insects on the wing, or the twitter of the solitary stone-chat, as it flits from rock to rock. The solitude is complete, the silence solemn and impressive. Our perfect isolation amid such a scene—the vast dimensions of the objects around us, and their expression of power, are true elements of the sublime, and awaken the most pleasing and elevating emotions. There is a delightful consciousness of a new activity in the fancy, and an increased buoyancy and intensity in the feelings. To the geologist there is another source of the sublime in contemplating the effects of the mighty forces which have rent the crust of the earth, raised these mountain masses from the fiery depths beneath, and scooped the glens and corries out of the solid rock.

28. The cliff on the north side of the corrie shows some interesting dikes. One of these is of green pitchstone, and cuts the granite sheer through in a N. and S. direction from bottom to top of the cliff. It is four feet wide, prismatic across, and owing to the more rapid disintegration, depressed below the level of the granite. The contact does not present any peculiar change in either rock, such as usually marks the plane of contact of pitchstone and the sedimentary strata. The pitchstone is decomposed into a thin white film in many places along the outer edge of the dike, next the granite, in consequence, probably, of the oxidation and removal of the iron which enters into its composition. The dike is in some parts of its course obscured by debris, but upon the whole is, perhaps, the best defined dike of this rock occurring anywhere in the granite of Arran. But phenomena of much greater interest will be again noticed in connection with other pitchstones.

Two basaltic dikes occur close together, about 100 yards east of the dike we have been describing; they traverse the granite precipice in the same manner, but in a different direction, their course being about 28° W. of N., subject, however, to undulations. These dikes are from eighteen inches to two feet broad, and are separated by a granite band eight or ten feet in breadth; elliptic masses of granite, of which the largest we observed was about eighteen inches by nine, are enclosed in the trap, but very little altered. The alteration, indeed, is nowhere remarkable, the granite being in some places coarse, in others fine-grained, along the planes of contact. Specimens may be obtained of both rocks firmly adhering.

Some pretty plants occur here in shady spots on the granite ledges, where a little soil has accumulated; they will reward the young botanist for his long walk to their secluded habitat. The *Sedum rhodiola*, *Oxyria reniformis*, *Saxifraga stellaris*, *Alchemilla alpina*, and several others, rejoice in the temperature and humidity which these heights supply.

To reach the head of the valley, and the first ascent of the Pass into Glen Iorsa, we now direct our steps along the base of the precipice, where the grassy tufts and granite debris afford a safe footing on the steep slope, keeping as high a level as possible, in order to shorten the ascent by which the summit is reached. As we pass along we notice several dikes of pitchstone and basalt, ranging north-westwards up the precipices on our left towards the summit of the ridge, and doubtless crossing down on the other side into Glen Iorsa; but we may not now delay to trace them. The Ben-Talshan ridge on our right, and that along which we have passed, coalesce at the head of the valley, and their union is marked by a very steep grassy slope, encumbered with granite blocks projecting from the soil *in situ*, or deeply imbedded in it, but free from the huge flat and smooth sheets along which it is difficult and dangerous to cross. This depression or break is in the direction of the head of the valley, and owes its origin to a basaltic dike, which appears at the beginning of the ascent, and is seen to enclose masses of granite, as in the case already mentioned. Its range is magnetic north and south, and width about twelve feet. The situation of the Pass to which we are now to mount, and which is fully 1,000 feet above us, is indicated by a bold rocky point, a little in advance of Beilach-an-id-bho, the last high summit of the Ben-Ghnuis range. Marking its position by the compass, and then pressing up the steep, we gain a wide and grand prospect from the summit of the Pass. Clambering to the top of the rocky point, we look down from a height of fully 2,000 feet into a nook or recess of Glen Rosa on one side, and into Glen Iorsa on the other. The descent towards the latter is easy; towards the former it should not be attempted; for, if practicable, it is highly dangerous. From the Garbh-Alt valley, in fact, the only access to Glen Rosa is by the way we have come up.

29. We are now at the southern extremity of the Ceims (Kyims) which link on Cior-Mhor to the Ben-Ghnuis range. This is the ridge whose sharp and rugged outline, seen from the shores of Brodick Bay, is well known as bearing a striking

resemblance to the profile of a distinguished living statesman and writer. The ridge is formed by the edges of vast tabular masses or sheets of granite, inclined towards Glen Iorsa at a considerable angle, and cut sharply down on the side next Glen Rosa, so as to present towards it a continued precipice, formed of successive tiers of granite sheets and rhombic blocks. The jagged outline is due in part to the irregular wearing of the coarse-grained granite, but still more to the intersection of the ridge by a series of whin dikes. The horizontally prismatic structure of these subjects them to a more rapid decay than even the friable coarse granite; and hence most of the deep notches of this jagged ridge mark the situations of whin dikes. The fact is curious and interesting, and has not been before noticed. Ranging up the front of the precipice from Glen Rosa these dikes cut right through the crest of the ridge, and pass down towards the fine-grained granite, which occupies most of the Iorsa valley, and rises up on the back of the Ben-Ghnuis range and the Ceims, as far as the level of many of the cols, that is, the lowest parts of the ridges between the glens. It seems highly probable, from the direction in which these and the other dikes already mentioned range, that they are the same as those which were found crossing the fine-grained granite tract on the west side of the Iorsa valley (Art. 6).

30. The jagged and notched character of the ridge makes it impossible for us to pass along it on our way to Cior-Mhor; neither can we safely cross the huge granite slabs at the back of the ridge, as they are smooth, slippery, and considerably inclined. But below these a safe, though rough and irregular, pathway will be found; and from this we can occasionally pass upwards towards the ridge, along the clefts in which the dikes lie, to have a peep down into Glen Rosa, or to scan the frowning cliffs on the north side of Ben-Talshan. For the latter purpose a telescope will be useful. We must be careful, however, to return by the cleft by which we came up, till we reach the path. The granite sheets must not be attempted, as a single false step upon them might hurl the climber with fearful

velocity into the valley at their base, filled with blocks and debris. The range, width, and structure of the dikes are well seen as we pass along; the width is various, in some five or six feet only, in others ten to twelve, and fifteen to eighteen feet. The broadest, being of this latter width, is one on the south side of the col, between Glen Rosa and Glen Iorsa. It consists of a crumbling greenstone, and ranges 29° W. of N. The others have various ranges between W. and N. Many small shining flakes of crystalline oxide of iron occur in the trap of these dikes. The ridge of the Ceims is composed of the coarse-grained granite; but the path at its west base, along which we have come, is partly on the fine-grained variety. This appears generally at the height of about 1,600 feet on the cols, and on the west side of the ridge, but is not seen in Glen Rosa nor on Goatfell. It disappears on the ascent of Cior-Mhor, and is succeeded by the coarse-grained variety. The contrast is remarkable. The rock has quite a different aspect, a different feel under the hammer, and a peculiar style of disintegration, giving smooth outlines, and an entire absence of the aiguille-like highly picturesque forms, into which the coarse variety is resolved by the action of the atmosphere. The decomposition of both rocks conceals the contact, and we were not so fortunate as to discover anywhere the actual junction of the two varieties.—But the day is waning, and we have yet to scale the lofty peak of Cior-Mhor, shooting grandly up 900 feet above the ridge on which we stand. Though right to the summit "we might press, and not a sigh our toil confess," we must pause now and again to mark the ever-changing features of the magnificent scene gradually opening towards the west and north, and the new aspects in which the rugged crest of the Ben-Ghnuis range now appears. We must note, too, the change in the rocky floor over which we are passing. We leave the fine-grained granite on gaining the foot of the steep ascent: thence to the summit the mountain is composed wholly of the coarser kind. It is disposed in irregular tabular masses, split up into rhombic or cuboidal forms by fissures, independent of disintegration, and coeval with the solidification of the rock. The thinner masses we have called sheets; in both, the divisional planes separating mass from mass, and the fissures perpendicular to them, are alike the result of crystallization on the large scale, and bear no analogy to stratification, which is the result of sedimentary deposit. The disintegration of granite, porphyry, and other igneous rocks, is mainly determined by these lines of separation; in some granites, but more remarkably in porphyry and the trap rocks, by a concretionary structure, which has resulted from the mode in which the crystalline centres of affinity develop themselves at the first parting of the heat of fusion, in a melted mass beginning to solidify. The schistose and prismatic forms, under which granite often appears, are but slight modifications of the forms already noticed, depending on the relative position of the divisional planes. The schistose form has often been described as a true stratification; but this structure is not continuous in one direction as strata are, nor does it exhibit the fracture or incurvation of beds; it is in truth but a local modification of the rhombic or cuboid form, under which granite more frequently appears.

31. The summit of Cior-Mhor, narrowed by distance into the form of an alpine aiguille, is found to be an irregular elongated plateau, large enough to accommodate a small picnic party. The rugged shoulders flanking the peak are huge ritted masses of bare rock, separated by clefts which descend far into the heart of the mountain. On three sides there are precipices, through clefts in some of which the top may be reached; an easy ascent is possible from the west side only. Thrown forward on a salient angle of the western ridge, and little more than 300 feet lower than Goatfell, Cior-Mhor affords a commanding view of the ridges and dividing valleys, the peaks and precipices of this singular mountain group. Its situation, as the geographical centre of the tract, has been noticed already, and the relations of the various ridges pointed out (Art. 2). Viewed from the summit where we now stand,

the scene is very wild and grand. The ridges swell up steeply and nobly in front of us from the very depths of the glens, in their majestic forms of "peril and pride," and stretch away on either hand, shooting up here and there into the highest peaks, and cut, in lower parts between, into spiry fantastic crests. The craggy precipices and long steep fronts of naked rock have an imposing expression of sternness and power. Crowning the ridge of which they form the lateral supports, Goatfell presents on this side its grandest aspect. The eye, from its elevation, takes in, under a large visual angle, the entire western steep, from the summit of the mountain to the bottom of the glens. On the north side of Glen Sannox the ridge of the Castles and Suithi-Fergus starts up with little less of suddenness and grandeur. Lying deep down at the foot of these lofty ridges, and closed in on the south-west by the high mountain on which we stand, Glen Sannox has an air of singular loneliness and solemnity. The same breadth of form and grand scale of parts are found nowhere else in Arran;—the very simplicity of the composition is one of the greatest charms of the glen. The silver thread of its river, meandering far out eastwards, leads to the world of life without; and the murmur of distant waters, stealing up from its sombre depths, breaks pleasingly the awful stillness of the summer day on these high peaks.

32. We have attempted, in preceding paragraphs, to indicate the successive steps of the process by which the mountain nucleus acquired its actual conformation. From our present commanding position we are better able to estimate the amount of elevatory force required to raise the high peaks and massive ridges around; and the length of time and intensity of erosive agents which the formation of the long and deep chasm dividing the ranges demands. Fill up this chasm and the other glens with solid granite masses, to the level of the peaks and ridges, over all throw a mantle of slate, continuous with the present circular boundary around the nucleus, depress the area full three thousand feet, till the ocean flows freely over all, and some measure will be obtained both of

the force and of the time through which the present aspects were assumed.

33. The sunbeam is the joy of this mountain wilderness. It lights up the solemn old rocks till they laugh into beauty under its bright spell. Though devoid of vegetation, which might throw bright tints around the rugged surfaces—for the saxifrage and alchemilla, the cryptogramma and other ferns, the club moss and juniper, nestle in shady clefts, and small patches of grass occur only here and there among the blocks—yet are not these bare rocky masses without a certain natural adaptation to produce warm harmonious colouring. The three ingredients of granite have peculiar shades and different reflective powers; oxide of iron, always present as a constituent, passes in decomposing through various rich tints, and the rocky surfaces themselves, smooth or rough, dry or moist, are often dotted with small lichens. The result is a sober but pleasing tint in keeping with the general expression of the mountain scenery; it runs through various shades of gray, purple, and a tempered red or orange. The effects seen from Cior-Mhor are finest in the afternoon. Marvellous contrasts now lie athwart the stony ridges and deep glens, adding a wondrous charm to the scenery. In the depths of Glen Rosa the sun has gone down an hour since, and a deep gloom has settled on the dark recesses of Glen Sannox. Sharp shadows of the western ridge, showing a perfect profile of its jagged crest, are slowly creeping up the western front of Goatfell, whose summit is bathed in a flood of glorious orange light. Thrown back from rock to rock in mellowed and harmonious tints, it maintains a bright twilight along the base of the western ridge, by which we must descend, and throughout the upper part of Glen Rosa, along which our after path will lie. Taking a last survey of the surrounding peaks in their gorgeous evening tints, and contrasting the bold rocky foregrounds, now flooded with light, with the smoother and fading outlines of the lower hills, we must hasten downwards. The far off landscape we shall see better another day from the summit of Goatfell.

The low ridge or col connecting the base of Cior-Mhor with the next height to the south, breaks down steeply, but without precipices, towards Glen Rosa. We can descend easily at almost any part of it. A few minutes will bring us to the junction of the two burns, over the most rough and toilsome part of the long walk yet before us. Farther our way lies nearly by the side of the Rosa burn, through the moss and heather, till we reach the Garbh-Alt, where our ascent began in the morning. The smooth carpet of the glen is now beneath our feet, and we dismiss all fears of adders in the path, which troubled us a little in the uncertain light of the last hour. Emerging from the glen, and crossing the pretty burn by the rustic bridge, we gain the Brodick side and Glen Shant lane. In twenty minutes more we are at the door of our comfortable inn, and are soon seated at the welcome evening meal. It is pleasant to talk over the incidents of our long walk; they will awaken joyful memories on wintry nights, years to come.

EXCURSION II.

THE CORRIEGILLS SHORE.

34. TO-DAY we shall stroll leisurely by the sea-side, and study the eruptive rocks which break through the Corriegills sandstone. The botany of the shore is rich, and the pools in the tide-way teem with life. We shall notice in this place the geological phenomena only. A group of whin dikes marks the first emergence of rock from beneath the Brodick sands; and from this point eastwards the rocky platform exhibits a complete network of interlacing veins and beds of igneous products, traversing the sandstone strata. Between the end of the sandy beach and the landing-pier at Spring-bank there are several cases of bifurcation of dikes, and of a singular crumpling of the sandstone strata. The bed of the burn within the pretty grounds connected with the hotel shows several greenstone dikes, and the beds of sandstone are well seen in the banks and rocky ledges over which the water tumbles. On the shore fine sandstone and conglomerate are irregularly intermixed, indicating periods of sudden and gradual deposit, and varying forces in the transporting currents. The sandy strata are red, yellow, and white, and, as we advance eastwards, predominate over those of conglomerate structure. The fragments in the conglomerate are mica slate, like that of Cantire, slate similar to that of the nucleus, white quartz, and quartz of that peculiar resinous variety (Art. 7), for which we know no locality in Arran. Pieces of porphyry also occur, but no fragments of granite. The whole series dips nearly south, at angles varying from 15° to 20°. The upper surface of the sandy beds is worn in a singular way, portions more quartzose, or with a calcareous cement, standing out in thin, sharp, irregular ridges, while

the parts around are worn away, being softer or more ironshot. The rock has thus a honeycomb structure, like what one often sees in the worn corallines of the mountain limestone. Numerous bowl-shaped cavities also occur, due, probably, to the removal of imbedded quartz balls, or the grinding action of these, by the movements of the waves, when once loosened from their bed in the sandstone.

Some general remarks have been already made (Art. 17) on the dikes of this coast, and we shall now only notice the individual cases of most interest. The whole number is about sixty, and the direction generally between N.W. and N.E., a few running nearly E. and W. Most of them alter the strata of sandstone more or less. The great majority are depressed below the level of the sandstone, owing to the more rapid disintegration depending on their structure, as already often noticed. They are of all widths, from eight inches to forty feet, and the sides are generally parallel, and the course rectilinear or slightly undulating. Most of them traverse quite across the rocky platform, and are continued into the cliffs, up whose front they are seen to range, either level with the surface, forming deep gashes, or projecting like walls. These cliffs are the old sea margin, and are hollowed into caves along their bases, and otherwise sea-worn to a considerable height. The gashes in the cliffs were doubtless formed when the sea stood higher; the process being now completely arrested in such situations. But dikes placed under circumstances exactly alike do not waste with the same rapidity. Though the prismatic structure is the same, the chemical composition varies, as does also the internal texture, while the adjoining sandstone varies also in its capability to resist decay. When the alteration produced by the dike is great, the sandstone will resist disintegration; if the contrary is the case, the sandstone may wear rapidly, and the dike project. "From some experiments made several years ago," says Mr James Napier (paper quoted in Art. 17), "on the decay of trap boulders, I found that certain varieties of that rock are rapidly changed by the action of water; lime and magnesia being dissolved out, the iron converted into a peroxide, and a crust formed on their surface, which is brittle and easily abraded." Mechanical and chemical differences have thus both to be considered, as well as the *relative* powers of resistance of the dike and the containing rock.

A remarkable group of dikes occurs under the east end of the high cliffs, near the point where the shore bends southwards. One of these is the broadest dike of greenstone on this coast. Its general breadth is twenty-five feet, but it widens at one place to forty feet. The sandstone is rendered very hard and quartzose to the distance of several feet. The range is 47° W. of N., and the inclination east at a small angle. A deep fissure marks the course of the dike up the front of the cliff. This dike is noticed by Playfair, in his Illustrations of the Huttonian Theory (*Works*, vol. i., Art. 266), as producing a marked change on the sandstone, and as indicating the relative durability of the two rocks. A little east of this dike is another, nine to twelve feet wide, inclined to the west, and ranging 18° E. of N. It offers no remarkable appearances; but the next dike east of it, though but seven feet wide, alters the sandstone more than any other of the whole series. This is probably owing to the nature of the rock, which, being a highly crystalline greenstone, must have passed slowly from a state of fusion. The stratification of the sandstone is obliterated through a space of seven or eight feet, and this rock assumes the structure of a claystone. The case is strongly in favour of the view often advanced, that the Arran claystones are merely metamorphic sandstones. Intersecting this dike is another, ten or twelve feet wide, ranging 37° W. of N., dipping E.N.E., and consisting of compact fine-grained greenstone. It is sunk below the level of the sandstone; and, on this worn, depressed surface, there rests a boulder of coarse-grained granite, estimated at about thirty tons weight. Now, no force of surging waves, surged they ever so fiercely, could shift the position of such a mass as this; and we must, therefore, conclude that the huge boulder now rests where it was originally thrown down by the glacier or floating berg, which bore it from the granite nucleus. Perhaps, however, it is not necessary to suppose that the dike was then excavated to its present level. It is quite conceivable that as the parts of the dike around the boulder were worn away, its support may have been loosened, and so its position may have shifted a little at long intervals.

35. In the sandstone of the cliffs overlooking this point we were so fortunate as to discover a true carboniferous fossil, a species of *orthoceras*—the occurrence of which in this locality is strongly in favour of the view already advanced (Arts. 8, 24), regarding the age of these sandstones. It may, indeed, be urged that carboniferous fossils pass upwards into the lower portion of the Permian (lower new red) system. But we submit that orthoceras is not a genus of which this can be said; a very few cephalopods, allied to nautilus, being the only animals of this order yet found in the lower Permian beds.

Still advancing eastwards, we meet with several other dikes. One of these is depressed ten feet in the tideway, is on a level with the sandstone along the grassy surface at the base of the cliffs, and in the front of the cliff is again worn, forming a deep chasm, which is bounded eastwards by a bold projecting edge of the cliff, crowned with wood. The dike on the beach, and fissure aloft, are about fourteen feet wide each, but not exactly in the same direction, the dike sustaining here a considerable undulation.

36. We now reach the great bed of claystone, the largest upon this coast, and presenting many interesting appearances. It forms a vein rather than a bed, as it is placed at an angle of about 15° with the sandstone strata. The sandstone dips 12° E. of S. at about 15°, while the claystone vein is inclined in the same direction at an angle of about 30°. The rock is divided irregularly into prisms by joints perpendicular to the lower surface of the vein, so that the prisms lean back towards the south, giving the appearance, when viewed casually, of a

bedding directed northwards. The structure at the upper surface is often schistose at right angles to the joints, or in the direction of the vein. The base is of a uniform texture, of felspathic substance, with quartz pieces imbedded. The structure varies from a uniform claystone, or clinkstone, to a small-grained porphyry. The colour is pale yellow, or yellowish-white; and at the first view the rock might be taken for a sandstone: it has indeed been described, when seen in its continuation in the adjoining cliff, as a white columnar sandstone (Headrick's *Arran*, p. 66). The rock on which it rests is a conglomerate; that beneath which it plunges southwards is fine-grained sandstone. The upper surface of the vein is very rough and jagged, with no resemblance to the style of decomposition among sandstones. The breadth exposed upon the shore is between thirty and forty yards. Along the level shore, where the vein rises to the north, its lower surface is exposed in a wide fissure hollowed out of the sandstone by the action of the sea. It is here seen to rest upon a vein of trap, three or four feet thick, and having the same inclination as the claystone. The line of contact is irregular, and in two or three places thin bands of conglomerate are interposed between the trap and the claystone. The lower portions of the claystone, next the trap, are harder, or converted into hornstone; the conglomerate is much indurated, and assumes the dark colour of the trap; while the latter becomes a fine basalt, and is intermixed with the sandstone below, or dispersed through it in lumps. The posterior origin of the trap vein is thus clearly indicated. The appearances are correctly described by Dr MacCulloch (*West. Isles*, vol. ii., p. 403), and an illustrative drawing given (vol. iii., plate xxiv., fig. 1). The sandstone overlying the claystone along the south side of the vein is very slightly altered.

Several dikes traverse the sandstone platform between the claystone vein and the great boulder, some running nearly N. and S., and others nearly E. and W. The former seem to shift the latter, producing a change in the direction of more than

20°. The dip of the sandstone is also affected by these dikes, being thrown round about 25° towards the west. One of the dikes, sunk more than two feet below the sandstone, and eight inches broad, is lost on entering the claystone: it may be connected with the underlying trap vein. Another, close to the boulder, six feet and a-half broad, consists in the centre of blue-coloured, rapidly decaying greenstone, and at the sides of a hard crystalline variety of the same rock, standing above the level of the central parts and of the adjoining sandstone.

37. The celebrated Corriegills boulder, under whose shadow we are now resting, is of imposing dimensions, and a conspicuous object from all parts of this coast. A few on the Corrie shore exceed it in size; but they are close on the edge of the granite nucleus, and we may suppose it quite possible that if Goatfell "shook his giant sides" under some earthquake throe, they might have been hurled headlong from the summit to the sea-level. Other causes must be sought for the transport of this enormous mass from the parent mountains; and of others still farther removed, though of lesser magnitude. We have already considered the only possible causes, and attempted to estimate the evidence in favour of each (Arts. 20, 22). That to which we chiefly lean receives support from the case before us. A crowd of lesser blocks surrounds the huge boulder of which we speak—an association much more likely to occur in the case of glaciers or bergs, than of currents emanating from a centre so remote. The cubical contents, and consequently the weight, are very difficult to estimate on account of the irregular form. The dimensions at the base are 21 feet by 12, and the height 15 feet. If rectangular, it would weigh 315 tons; but if we deduct one-third for the conical form, which is a large allowance, we shall have a weight of 210 tons.

The cliff we have had on our right all along subsides here, and the cultivated fields of the Dun-an and Corriegills farms come down to the water's edge. The northern mountains are hidden from us, but the picturesque ridge of Dun-fion sweeps finely round, abutting on the sea in a lofty cliff. The farm-houses nestle cozily in against the hill sides, sheltered from winter storms, and forming a delightful summer retreat for those who love retirement and quiet simple beauty. The grander features of Arran can be seen in the distance of an easy walk.

38. South of the boat station, under the farm-house, the dikes traverse the sandstone in every possible way; intersecting one another at various angles, bifurcating, lesser ones lost in larger, etc. One of them is exposed through a longer course than any other dike on the coast. It runs a long way parallel to the line of the shore, or almost due N. and S., till lost under the sea near Clachland Point. It is fourteen feet wide, sunk under the sandstone in the tideway, and sends off a branch towards the N.W., a tongue of altered sandstone being at the bifurcation.

Two pitchstone veins, one of claystone, and one of quartziferous porphyry, are found on this part of the shore. The lesser pitchstone vein traverses the level shore obliquely about half-way between the boat station and the base of Dun-fion. Within the tideway it ranges about 72° W. of N.; then bends about 28° towards the S., *i.e.*, runs about due W., and bends again into the former course before it enters the sandstone cliff. Under high-water mark it is about three feet wide, in other parts five or six feet. Seaward it is placed conformably among the sandstone strata, and is irregular in direction and breadth. A dike of greenstone here cuts it nearly at right angles, but the appearances are noway remarkable. This vein, or bed, has been overlooked by MacCulloch and succeeding observers. The greater pitchstone vein is conspicuous, forming a broad band in the front of the sandstone cliff farther S., and has been noticed by every one. It occupies a slanting position in the cliff, parallel to the sandstone strata, dipping with them towards the S.S.W. at nearly 30°, and rising towards the N.W. In the opposite direction, or towards the S.E., it seems to plunge beneath the sea; but the debris here obscures it, so that its course cannot be traced to the water's edge. Climbing up the cliff to examine it more closely, we find it to be 13 feet 5½ inches in thickness; of lamellar structure and dark bottle-green colour. There is no remarkable change on the sandstone—a slight induration merely, but the lower portion of the pitchstone is changed into a blue-coloured, porous, slag-like matter, like a pumiceous lava: this, however, is probably the mere result of decomposition. Its exact position among the sandstone strata is not easily determined: if parallel to them we must call it a bed, if intersecting them, a vein. Fallen masses of the pitchstone strew the beach; and among these, where the path comes close on the water, there is a dike of red quartziferous porphyry, nearly perpendicular to the shore, twelve or fourteen feet broad seawards, but narrowing inland to five or six feet.

A granite boulder, about one-third of the size of the one at Dun-an farm-house, rests here in the tideway; and great numbers strew the shore all along.

39. Between the two veins of pitchstone which we have just described, there occurs upon the flat shore another large vein of claystone remarkable for a peculiar structure. This is developed in those parts only which are near the junction with the sandstone, along the south side of the vein. The appearances are fully and accurately described by Dr MacCulloch (vol. ii., p. 405). The structure referred to "is concretionary globular, or striated; the latter being either found separate, or united with the globular in the same specimen. The former puts on sometimes the appearance of spots, circular or elliptic, resembling Siberian jasper. The spots, as well as the stripes, are attended with corresponding differences of hardness, the former arising from the globular structure, the latter from a schistose or laminar one. The spots being often elliptical, compressed, or elongated, occasionally become laminæ in the progress of elongation, passing into them by insensible degrees. The vein next the sandstone varies much in hardness, but it cannot be said that this induration bears any relation to its proximity to the sandstone. The concretionary structure is seen both in

the hard and soft varieties, it is radiated fibrous, the radii sometimes diverging from a point, and sometimes from a solid nucleus, which is further, in some instances, surrounded by a white earthy crust. In the progress of induration the rock at length loses its character entirely, appearing to pass into a substance of an indefinable nature, of a horny aspect and dark dull green colour—partaking of the character both of calcedony and pitchstone. It has been described as globular pitchstone; but it is far removed from this rock by its extreme toughness, want of lustre, and by the form of its fracture." Mineralogists have long regarded this curious rock with much interest, and various opinions have been held respecting its true relations, some considering it as allied to claystone, and others to pitchstone. A comparison of the appearances observed here with those seen at Tormore, and in Moneymore glen, has led us to conclude, that "the substance of an indefinable nature" is hornstone; and a transitional state of these earthy matters, between claystone and pitchstone; the globules being due to the formation, within the mass when in a fluid state, of those crystalline centres of radiation, which determined the prismatic and jointed structure of igneous rocks, in the manner described in Mr Gregory Watt's *Experiments* (*Phil. Trans.*, 1804, *See* Excursion IV). The radiations are of pure quartz, while the base is of the same substance, but mixed with colouring matter, iron or manganese, and other trifling impurities. We shall again notice similar varieties in other places.

40. Near the pitchstone vein the high ridge of Dun-fion reaches the coast; and the trap which forms its summit appears in section in the cliffs, overlying the sandstone. It presents a façade of imperfect columns, and contrasts strongly with the stratified sandstone. Both sink rapidly southward in the direction of the dip; the sandstone is depressed below the sea-level, and the trap then occupies the coast, forming the low point at the entrance of Lamlash bay, called Clachland Point, and also Hamilton Rock, the small island in the chan-

nel. Near the point where the sandstone disappears, some remarkable effects of trap dikes are exhibited. On the west side of the point the sandstone suddenly emerges again, and extends to Lamlash. The whole bay, indeed, has been excavated in this red rock, which forms the base of the cone of Holy Isle, as well as the coast on the mainland. But we turn here meanwhile, and direct our steps to Brodick, by the summit of the ridge which divides the two bays. Its southern side forms a long grassy slope of gentle inclination towards Lamlash bay, on which are many granite blocks, and a few of felspar porphyry; towards the north it falls suddenly in steep cliffs and terraces. As we pass up the easy ascent striking views open southwards. Starting suddenly from the water's edge to the height of 1,020 feet, Holy Isle, with its encircling sea-line, fills the foreground grandly. The bay, with its fine double sweep to King's Cross Point, its wooded banks, debouching glens, and background of dark hills, forms the right of the picture; to the left the glassy sea sleeps in the sunlight, dotted with small white moving specks, and bounded by the winding line of the Ayrshire coast, with promontory, creek, and bay, along which the eye may range from Ardrossan heights to the Mull of Galloway. The scene is singularly sweet and picturesque; without Holy Isle as an integrant part it would want character: this stamps upon it peculiar features. The northern mountains, so essential and expressive in most Arran landscapes, are here hidden from us; but when we gain the highest edge of the ridge, they burst upon us with startling suddenness in an aspect quite new. They are grouped in a way not seen from any other point, and their jagged profiles are thrown into lines of singular boldness. Not less new and striking is the aspect of the lovely bay of Brodick, with its noble castle and hanging woods, of the glens into whose far depths the eye can reach, and of the "cottage homes" nestling amid groups of trees, in shelter of the hills and sloping banks which enclose the smiling fields of Brodick plain. Often as we have come to the edge of the ridge by this route, we have always felt the same delightful surprise when the scene first burst upon us.

41. This ridge consists chiefly of sandstone. Trap rocks form a thin capping along its highest part; they extend a little way down the southern slope, thickening as they descend, but along the highest ridge are so thin that sandstone occupies in some places depressions in the ridge; and the trap occurs only on the isolated tabular knolls into which the ridge is cut up, especially towards the west. The trap consists of felspar and augite mainly, with imbedded zeolites. MacCulloch designates it augite rock; but hornblende and iron also occur. The sandstone close to its junction with the greenstone of the summit is highly metamorphic, resembling a quartz rock. Along the steep northern front also, west of Dun-Dhu, beds of clinkstone crop out in various places, at different heights. These are either of truly igneous origin, or are metamorphic sandstones, altered by whin dikes, of which there are several, or by the near proximity of the great igneous masses—greenstones, porphyries, and pitchstone—which pervade the sandstone in this quarter. The association of these various products affords one of the most interesting sections to be met with in Arran. It is best exhibited a little to the west of the path which leads down from Dun-fion to the farm-houses. Dun-fion, or Fingal's Fort, is the highest point on this portion of the ridge, right over Corriegills shore, and 500 to 600 feet in height. A low mound, enclosing an elliptic space 40 yards by 16, is seen round the summit, but nothing whatever is known of the history of the fort. From traces of vitrification said to have been noticed on the stones of the mound, it has been conjectured that the place was the site of a beacon fire, in the wild old times, as well as a stronghold. We were unable, however, to find the least trace of such an effect; but certainly the situation was well chosen if such was the purpose. It looked out far and wide across the waters of the frith, so that no sail could pass or approach unseen; while the glare of its fires would light up every glen and hill-side on this side the island,

from Dippen to North Sannox.* Dun-Dhu is a prominent hill, nearly as high, a little farther west, and standing out in front of the cliffs, to whose base it is joined below. This hill is composed of columnar felspar porphyry; and between it and the path descending from Dun-fion the beds represented in the annexed diagram occur in a well-marked vertical section; two veins of pitchstone and one of porphyry, with sandstone intervening, surmounted by augitic trap.

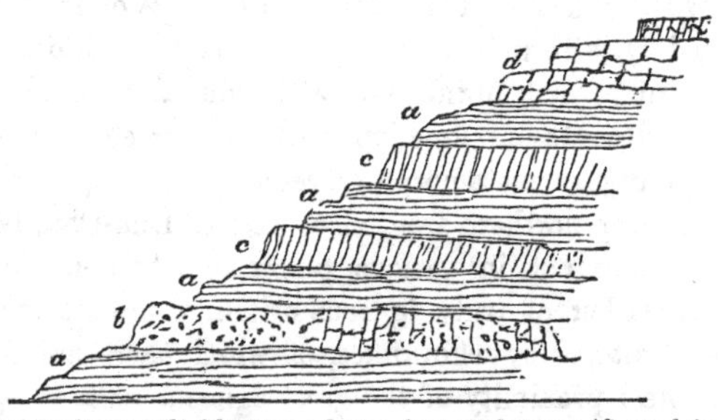

(a a a a) Sandstone; (b) felspar porphyry; (c c) pitchstone; (d) overlying trap.

The upper pitchstone vein, *d*, is about thirteen feet thick, and dips S.W., at a small angle, probably 10° or 15°, and may, therefore, be conformable to the sandstone which is seen in the quarry near the footpath to dip about S.W. at 10°. The rock is of a dark green colour, without felspar spots, divided into beds from eight inches to three feet in thickness; and these beds consist of closely aggregated prisms, or laminæ splitting into prisms. The bed is very high on the cliff, coming close up to the prismatic greenstone, but apparently separated everywhere by altered sandstone. Its upper surface is on a level with some of the depressions in the edge of the trap ridge; and it is apparently through the most eastern of these that this bed passes across to the back of the ridge, and is again seen behind Dun-fion, where a bed of sandstone is also seen. It no doubt passes downwards towards the bay, but it is not seen again: perhaps the trap of the cliffs is a

* Headrick's Arran, pp. 64, 76; M'Arthur's Antiquities of Arran, p. 90.

mere facing, and the pitchstone may traverse the sandstone only. It is the upper surface of the bed that is here exposed.

The lower pitchstone vein, *c*, is 14 feet thick, and of the same colour and structure as the other; but its low level can hardly admit its passing across the ridge. The vein of porphyry occupies a ledge about fifty yards lower down: this ledge being on the level of the front of Dun-Dhu, from which the porphyry extends. The sandstone strata are not well seen on the ledges *a a a*, between the beds, in the line of section; but come out distinctly on their continuations east and west.

Neither of these pitchstone beds has any connection with those upon the shore, which are far below the level of the base of our section. But the upper bed probably passes westwards across the hollows south of Dun-Dhu, since a bed of pitchstone, about twelve feet wide, again appears on the west bank of the first burn N.W. of this hill, and in a position to which a slight undulation in the upper bed would bring it. A few hundred yards farther west, in a field within the new fence, it breaks out again with a breadth of more than twenty-five feet, and, undulating again, passes up westwards by the crest of the moor, where it appears of considerable breadth, there being only indications of its occurrence between these two points. But from the crest of the moor it can be traced continuously into Birk Glen. (See Excursion IV.) The rock is most usually glassy, of a bottle-green colour, homogeneous, and very seldom exhibits any porphyritic structure. Between the two first outbreaks, on the bank of a stream, there is a bed or dike of a very hard and tough hornstone, or brownish grey quartzite, with imbedded bits of quartz, a slight modification merely of the pitchstone base; there is a greenstone dike beside it. The relations of these igneous products are very curious.

42. The northern front of Dun-Dhu consists of fine columns of porphyry, divided at irregular intervals by flat joints. On the N.W. side the columns are very perfect, and have a singular diverging fan-shaped arrangement. We ascend the

hill easily from the south side, and find the summit composed of huge prismatic masses of porphyry, lying closely side by side almost in a horizontal position, and in a direction from N.E. to S.W., so that their ends, cut uniformly off, form a wall towards the west, ten to fifteen feet in height. Some of these columns are twenty feet in length, apparently without joints: in general, however, there is an indication of a flat jointing, at distances of four to seven feet. Most of the prisms are four and five-sided. The rock has a gray felspar base, occasionally ironshot throughout, or merely streaked with iron; and the imbedded crystals are of glassy felspar. Bits of quartz are also disseminated through the base; apparently minute crystals with their angles rounded off, as if by attrition.

No granite blocks were noticed on this detached summit; but many of considerable size, as much as from six to fourteen tons, were observed close to the cliffs on the west side. They are found in great numbers, and of large size, over all parts of the open plateau between the base of this ridge and the shores of Brodick Bay. They have been mostly blasted hereabouts, gathered off the fields, and reared into fences. Many of them are rounded, polished, and marked with very perfect glacial striæ—traced upon them ere yet detached from their native beds in the granite nucleus, or else in their transit upon the ice to their present resting-place.—Across this plateau our path homeward is by the tortuous lane which enters on the Lamlash road immediately below the Free Church. It pursues its winding way between fragrant hedge-rows, through birky places by burns, past cottages with their corn-fields and meadows, and affords besides, in its distant outlook, much to beguile the way. The evening tints of rich purple-gray on the northern mountains, beyond which the sun is now going down, and the profile of the rugged peaks against the golden sky behind, are glorious to look upon; the light is strong yet on the peaks of Ben-Yim and the Cobbler, on Benlomond and the mountains of Aberfoyle; the broad shadow of Maoldon falls dark athwart the sea; but beyond this a flood of light comes streaming down the northern channel, and brings out strongly the gray rocks of Garroch head and the Cumbrays, and the lovely glades amid the woods of Fairlie. The turrets of Brodick Castle are yet gilded by the sunlight which falls in broad sheets across the lovely bay, whose glassy expanse is crowded with fishing parties. Before we can reach our stately but pleasant inn, the steamer has rounded Merkland point, disappeared in the shadow of the hills, emerged again into the "lanes of light," and threaded her way through the fleet of boats.—Already the fair throng has left the landing-place, scattered now in gleeful groups along the various roads.

EXCURSION III.

To the Summit of Goatfell.

43. Goatfell is an unmeaning corruption of the native name of this mountain. Gaoth (gāŏ) is the Gaelic word for wind; this may be the origin of the first part of the name; with the animal indicated the mountain has no sort of connection. Then, Fell is not a Scottish word; it belongs to the North of England, and to Scandinavia in its form of fjeld or field, applied appropriately to the wide flat mountain plateaux of South Norway. Bein, or with the aspirate Bhein (ban or ben, ven), is a mountain. Pen is the English form, as in Penyghent, Pendle hill, and hence the Latin term Pennine for a principal range. Thus the name would be Gaoth-bhein, or Bein-gaoth—the hill of the winds—not very expressive or special as regards this hill more than others standing prominent. But those who gave the name perhaps knew no higher hill; and there is a peculiar effect often seen here to which the name may perhaps have reference. The Ben-Ghnuis or western ridge first arrests the vapours ascending from the western sea, and condenses them along its winding summit into a dark sinuous bank, from whose shattered edge masses float away when the breeze gets up, and dashing against the flat side of the ridge of Goatfell, are driven in rapid eddies round its south end or over its upper edge. To one looking up from the quiet depths below, this would suggest the existence of a furious gale upon the summit. As the weather thickens and the clouds accumulate, the cone gets completely hidden, and the rolling vapours pass even lower than the mill-dam, veiling the edge of the great waterfall, which then seems to issue directly from the clouds.

But such a day as this will not suit for our walk; we must

wait for bright skies and still air—few walks in Arran will then please us more. Every lover of mountains has a keen desire to climb the highest summit within his reach, and many will choose this walk for their first excursion.

The woods which stretch westwards from the castle are crossed by two paths, by either of which we may pass upwards into the moors. The readiest way is by the new approach to the castle, which is not prohibited, and which we enter by the gate near the bridge over the Rosa burn. We keep on the avenue as far as the west end of the stone bridge, then turn to the left through an iron gate; and following the path a little past the wooden bridge, again turn to the left, and soon emerge upon the moor. The entrance to the other path, which it is preferred that strangers should take, is by the castle stables and old hotel, more than half-a-mile N.E. of the Rosa burn bridge. This meets the other path before it opens on the moor. Or from the mouth of Glen Rosa, reached as on our First Excursion by Glen Shirag, we can easily climb up the eastern flank of Glen Shant Rock. A path lies near the east bank of the Cnocan burn, as far as the mill-dam, where this stream is gathered from many heads. Beyond this a track is marked out among the granite blocks to the east shoulder of the mountain; thence, along the edge of the ridge, a rugged path conducts us to the summit, over huge masses of rock, and along the edge of Cyclopean walls. But the geologist has much to see on the ascent which this route will not shew him, and we must conduct him by another way.

44. Beds of fossiliferous limestone occur in the woods N.W. of Brodick Castle; they belong, of course, to the carboniferous system, and are higher in the series than the sandstones, which appear in the bed and banks of the burn before it enters the wood. They are regarded by Murchison and Sedgwick as an upthrow of the Corrie beds, a theory to which we do not object; but the facts may be explained otherwise (Art. 13). These sandstones are succeeded by the Old Red as we ascend the burn above the wood, the junction being somewhere about

the place where the wall enclosing the new plantation abuts against the bank of the stream. But the contact is not seen, nor is there any gradation visible. The Old Red is here a hard quartzose slaty sandstone, with many thin brown laminæ, and elliptic blue or white claygalls: the dip is back against the slate of the mountain, contrary to that of the overlying carboniferous beds, at angles varying from 55° to 70°. As we advance the strata become much obscured by debris; and a little above the point where the west-burn enters the main stream the dip is reversed or towards S.S.E., at about 65° to 70°. The rock is here coarser, and contains imbedded masses of resinous quartz and dark blue slate; farther on it is darker and finer, with claygalls, and the dip again appears to be towards the slate, but the stratification is obscure. In front of the first waterfall, amid a group of birch trees, a mass of quartziferous hornstone porphyry lies across the bed of the stream, but seems to terminate against the bank on the west side; on the east side its extension cannot be traced, so that most probably it is not a dike. It is a beautiful rock, and if readily obtained in quantity might be used for ornamental purposes. The base is a dark reddish hornstone, containing crystals of glassy felspar, and round bits of quartz. There are beds of conglomerate here; and the strata lean toward the slate, but less "end on;" the dip being about N. 4° W.

We are now approaching the junction of the old red sandstone with the slate; but as this is not well seen in the bed of the main stream, it will be more instructive to diverge to the left, up the course of the west branch, as far as the dark brow of slate, where the hill suddenly rises. Passing up the bed of this west-burn, we find the common red rock and conglomerate succeeded by flinty or quartzose sandstone, obviously metamorphic. The cause of this change is soon discovered. At a waterfall on the burn there is an outburst of a peculiar granite amid the sandstone strata. This is an intimate mixture of quartz and felspar, without mica; in fact, the Eurite or

Weiss-stein of mineralogists. It extends for many yards in the bed and banks—how far cannot be determined. Between it and the rocky brow, which is formed of the common dark slate, there are various metamorphic beds, conducting us by insensible gradations into the true slate. Some of these are white and gray flinty slates, others fine-grained, hard sandstones. It is thus difficult to decide to which series the beds ought to be referred. The strata have been assimilated by the metamorphic action to which both series have been alike subjected. A similar case occurs at the junction on the White Water, to be noticed on another Excursion.

Returning now to the bed of the main stream, and entering it some way above the waterfall at which we turned off, we find similar white and gray quartzites about fifty yards in front of a deep chasm, with a waterfall, cut out of the slate rock. The sandstone is not seen in close proximity; but the junction must be near, as this is the first point at which the slate appears in the bed or banks. Two dikes of greenstone, from opposite sides of the pool, unite at the edge of the fall, twice bifurcate, and twice unite again, enclosing two long elliptic masses of altered slate, and then continue as one dike right up the chasm. The branches are three to four feet wide; and the joint breadth, as one dike, eight or ten. The dike undulates in its course from 30° to 40°, conforming to the course of the chasm. The chasm is, in fact, due to the dike; an original depression, produced by a fault or the irruption of the dike, determined the channel for the stream, and along the course of the dike the water met with least resistance in its work of disintegration. The chasm is nearly half-a-mile in length, with perpendicular walls ten to fifteen feet high, above which the banks rise very steeply on both sides. It runs in against the great sheets of slate, forming the waterfall below the mill-dam. At the base of these sheets the dike is seen again, interrupted or broken off in one place by the slate, from beneath which it again emerges, and appears upon the high brows above in the bed of a small stream entering from the N.E.

45. We are now at the famous junction of granite and slate close to the mill-dam. The appearances have been often described; we shall quote the very clear account given by Professor Ramsay (*Arran*, p. 4):—"The absolute junction of the two rocks is not here visible; but that it is in the immediate neighbourhood, probably in the bed of the dam, is clearly shown by the appearance of a granite vein, about one foot broad, which penetrates the strata, and crosses the bed of the stream about ten yards below the artificial wall which confines the water of the dam; thus indicating its intrusion, while in a state of fusion, into the stratified deposit with which it came in contact. The granite is of a yellowish colour, fine-grained and compact in texture, and consists principally of felspar. The slate is exceedingly tortuous; and the strata are intermingled with numerous veins of quartz of varying sizes, and which generally alternate with the slaty strata in regular minute laminæ." This description is correct and well stated; but an important fact has escaped notice altogether. It has not, indeed, been alluded to by any one of the many observers who have visited this locality, owing probably to the state of the water in the river at the time when this junction was examined. We refer to a dike of greenstone which crosses the river diagonally, ranging about 20° E. of N., and about thirty feet wide. It enters the east bank of the stream under the mound or wall of the dam, and is seen again on the surface, a little way toward the N.E., but is soon lost under the heaps of granite blocks. The strata of slate range 65° E. of N., or almost E. and W. by the compass, dipping 25° E. of S., at an angle of 70°. Thus the direction of the dike makes an angle of 40° with that of the slate. Now, there are granite veins in this dike; and these cross out into the slate on the east side of the dike, continuously without a change of direction. There is no mistaking the rocks, both the greenstone and slate are perfectly well marked; a portion of the former, traversed by the veins, has distinct acicular crystals of white felspar, and that concentric

structure so peculiar to this rock. The slate is quite homogeneous—a dark blue hard quartzite, in some specimens resembling Lydian stone—but the alteration from its ordinary state is not remarkable. The principal granite vein is several inches (six to eight) wide, and is traceable in the greenstone through a space of fifteen or twenty yards; and runs right on into the slate. Several smaller veins ramify through the greenstone, dividing and then uniting again. These have the structure so well described in the above extract. We conclude that the slate was first injected by the greenstone dike; and that the irruption of the granite, at a subsequent period, pierced through both of these rocks. The case is very interesting, being the only one in Arran, that we know of, in which greenstone is proved to have been erupted anterior to the injection of the granite amid the strata of slate.

46. We recommend the geologist, who is not deterred by the prospect of "a pretty stiff pull," to mount at once by the southern shoulder of Goatfell, avoiding the common pathway, already pointed out, which presents much less to interest him. He will thus have an opportunity of examining those huge natural ramparts of granite blocks, piled mass on mass to a great height, like the Cyclopean walls of Tadmor or Heliopolis, and of studying the structure of a granite mountain, the varieties of the rock, and its peculiar style of decomposition. We especially recommend this course if he has not accompanied us on our First Excursion.

We have now reached the south summit of Goatfell, the highest point in the island, elevated 2,865 feet above half-tide level. The north summit is 237 feet lower. There is not, perhaps, in Scotland, another mountain peak which looks abroad upon a scene combining the same variety of grand features. Many afford wider and finer mountain views; here the eye ranges over a vast extent of broken coast, the whole expanse of the noble frith, and its many narrow branches winding far in amid mountain solitudes. How grand from this commanding height are the surrounding peaks and rugged ridges, and the profound dividing glens! what lovely pictures in their glassy frames are these sister islands! how stirring the rapid movements of life all day on the inner frith! what a world of human interest in that great ocean steamer starting on her outward voyage! The "sweep of the circling horizon" embraces a magnificent amphitheatre, reaching from the mountains of Donegal and Londonderry on the W. and S.W. to Benlomond and Ben-Ledi on the N.E.; from Ben-Nevis and the mountains of Mull on the N. and N.W., to the ranges of the South Highlands, the Mull of Galloway and Isle of Man, in the opposite direction.

On a clear day, with a N. or N.E. wind, the panorama of the northern mountains is very fine. On the N.E. horizon, Benlomond is easily known by its advanced position, elongated form, and double top—west of it on either side may be seen Ben-Voirlich (Loch Earn), Ben-Ledi, and perhaps Ben-Lawers; north of it is the group of Loch Voil and Loch Dochart, among which Ben-More is conspicuous by its conical form. To the N.W. of Benlomond is the Arrochar and Loch-an-Slui group, among which Ben-Voirlich, the Cobbler, and Ben-Yim are the most conspicuous, the last farthest to the N.W., of an elongated form like Benlomond, but higher, and with three tops. Still farther round to the N.W. is the lofty group of Tyndrum, and Breadalbane forest, among which, perhaps, Ben-Lui may be recognised, with a deep corrie on its S.E. side filled with snow till far on in the summer. The wide extent of open undulating country, without mountains, between Loch Fyne and the Sound of Kerrera, and bounded N. by Loch Etive, renders it easy to identify Ben-Cruachan on the E. side, with its double summit, and Ben-More in Mull on the W. Ben-Nevis is less easily found; but having seen Goatfell from it on a very favourable day, we are quite sure it can be seen from Goatfell. It lies due N., far out on the horizon, and a little to the W. of Ben-Cruachan. The high group N.E. of Mull is that which lies between Corran Ferry and Strontian; N. of this are three conical mountains belonging to the district

W. of the "Great Glen of Scotland," somewhere near Loch Lochy. Carrying the eye southward, we see the S. promontories of Mull, with some low-lying isles, perhaps Staffa or Coll, across the north end of Jura, whose three paps are conspicuous and close at hand. South of Jura is Islay, well seen across Cantire, whose two dependent islands, Davar at the mouth of Campbelton harbour, and Sanda off the Mull, are under our feet. Across the top of Ben-Ghnuis is seen Rathlin Island, south of which rises the lofty ridge of Antrim, of which the most conspicuous summits are Knocklayde, Aura, and Trostan, nearly 2,000 feet in height; in the centre of Antrim Slemish is remarkable by its isolated position, and its form, a truncated cone. The bold promontories of the north coast come out in succession towards the west, Bengore, Magilligan, and Malin Head; the top of the chain of Londonderry bounds the view in that direction. Far out southwards the eye may sometimes penetrate as far as the Isle of Man, and the groups of Morne and Skiddaw. In that direction, twenty miles S. of where we stand, the grand cone of Ailsa rises abruptly from the sea; beyond, the eye ranges far out S.E. and S. along the noble sweep of the Ayr and Wigton coasts with their fine background of mountains. As the day verges towards evening, a purple curtain slowly falls over the scene to the south and east, while the light is still strong on all the northern mountains. Towards the west the islands stand grandly up in a sea of molten gold, which, rising far out westwards, blends with masses of gorgeous clouds "set on fire with redness"—later still—

> The sun, descending,
> Leaves upon the level water
> One long track and trail of splendour.

Of mere geological interest there is not much on the summit of Goatfell. Bits of greenstone lie about as if a vein existed near. Veins of fine granite in the coarse variety, of which the mountain consists, have been mistaken for veins of claystone. These veins are clearly cotemporaneous, and have no connection with the fine-grained granite of the interior; it is too distant and too deep to have sent such fine veins and threads to this height. Dr MacCulloch noticed a magnetic property in the granite, which he afterwards confirmed by observations made on other mountains in Scotland (*West. Isles*, ii. 351; *Geol. Trans.*, 1st Series, vol. ii. p. 430, and vol. iv. p. 124). The greatest deviation of the needle which we noticed was on a block N. of the cairn, where it pointed to 75° W. instead of 26° W., the present variation; a deviation of 49° from the true position. Such being the irregularity, it may be well to state that the tongue of land between Loch Fyne and the Kyles of Bute bears due N. from the summit of Goatfell. Ailsa Craig nearly one point east of south.

47. If the day is calm and clear, one is disposed to linger on the summit and take in the full beauties of the scene. The Cyclopean walls on the north front are a fine study, shewing

the structure of a granite mountain, and the style in which the rock decomposes. About twenty courses of huge slabs are piled up into a lofty mural precipice. A walk is recommended down the N.W. front to the dividing ridge, past the north summit, to have a better view of the corries, and a peep into the dark depths of Glen Sannox. The first col on our descent is marked by a dike of gray trap about twenty-five feet wide, with felspar crystals, giving it a porphyritic structure, whose disintegration has determined the formation of the col. It cuts the Ceims, on the opposite side of Glen Rosa, and is there marked by a depression; the range is about W. 25° S. Another dike ten feet wide, with the same range, is seen a little N. of the N. top of Goatfell; continued across the glen, its course is right up the eastern front of Cior-Mhor. The wild grand cliffs along here are formed of huge sheets rising tier on tier towards the west, and dipping gently back east in the direction of the rims of the corries. We might easily descend by this route to the Sannox road, either by the S. base of Cioch-na-Oich, or farther S. by the hill-sides over the village of Corrie, to meet the evening boat for Brodick. But we must return back by the summit of Goatfell to examine the corries and descend another way, more picturesque and full of instruction. We have already attempted to account for the origin of the glens (Art. 32). Denudation by water, acting along fractures when the land was rising from beneath the sea, seems the most probable cause. Some geologists would ascribe their origin wholly to scooping out by ice; M. Necker to their being in the line of great dikes, along which, as in many cases already cited by us, disintegration would be more rapid. Whatever the cause, there must have been some original difference of surface to determine the first action. The origin of the corries is still more difficult to account for. One can easily imagine, that if such a bowl-shaped hollow existed when the snow and ice began to become permanent on the summits, ice should depend in great sheets from the rim, and work all round on the rock to its disintegration and removal, so as gradually to widen the corrie. But what gave the rim in the first instance? It must have been of considerable size to bring into play this supposed ice action, much larger than any hollow that could have been formed like those on the Corriegills shore, already alluded to (Art. 34) as excavated by the constant rolling about of large stones: so small a depression could scarcely determine the beginning of ice action. Volcanic action has been suggested; and certainly some of the corries are not unlike volcanic craters breached at one side. But it must be remembered that granite, though once in a molten state, was not erupted in volcanic fashion—the matter of granite so erupted would become pumice or obsidian, or mayhap, if toward the base of a cone, pitchstone or basalt. Volcanic action is thus inadmissible, and we can only conclude that the same cause that removed the mantling slate from over the granite (Art. 13), and began to mould the peaks, may have determined the first formation of the hollows; and ice, as in Alpine tracts, may have done the rest.

The descent from the summit of Goatfell into Glen Rosa and by the Rosa col must be carefully selected, as it is very dangerous in some places.—A few years ago a young man lost his life by a fall on this descent; and, more lately, a gentleman from London, left behind by his party, got into such a position that he could neither descend nor mount, and had to remain in durance over night, till released next morning by a party of men sent from Brodick to search for him.—For the geologist, two descents are more instructive than the rest. One of these is eastwards by Maoldon and the White Water, but it implies rather more fatigue than most persons will consider consistent with enjoyment, and we shall, therefore, reserve for another Excursion an account of what is to be seen on this route. The other is by the plateau S.W. of the Mill-dam, along the boundary of the granite and slate, where some fine junctions, granite veins, and contorted schist may be studied. The line of junction descends the N.W. front of Glen Shant Rock towards the mouth of the Garbh-Alt in Rosa burn, by which we passed up the opposite hill in our First Excursion. On the edge of the plateau, a quarter of a mile S.W. of the mill-dam, a good example of a "roche moutonnée" may be seen. The exposed horizontal surface is fourteen or fifteen yards in extent, rounded neatly off diagonally to the laminæ; the north-west front, or vertical face, is also ground off, but the opposite face turned S.E. retains its original form. This indicates the direction of the abrading force.—The whole surface of this plateau ought to be carefully searched for glacial striation, and other examples of "roches moutonnées."

EXCURSION IV.

The Wayside Museum.

48. The student of geology will make but little progress in this delightful science without a thorough practical acquaintance with the mineralogical distinctions of rocks. He must first know a few of the common minerals—quartz and its varieties, as rock-crystal, amethyst or purple quartz, smoke quartz or cairngorm, hornstone, jasper and pitchstone; felspar and its varieties, as glassy and compact felspar and albite; mica, talc, chlorite, serpentine, hornblende, hypersthene, augite, olivine, and calcspar. All these are "simple minerals" to the mineralogist; the chemist calls them "compound," since he can resolve them into elements. Using, then, the terms simple and compound, in the mineralogical sense, we divide the rocks into two classes; common slate, limestone and the marbles, coal, quartz rock, and serpentine, are the simple rocks; all others, forming by far the greatest proportion of the mountain rocks of the globe, are composed of two or more of the five simple minerals—quartz, felspar, mica, hornblende, and augite. Knowing these, the student will easily make out the composition of almost all the rocks, certainly all those of Arran. With every term a definite idea will then be connected, and the study of the subject will be smooth and pleasant henceforward. Arran has been called an epitome of the globe; our Wayside Museum is an epitome of Arran, in so far as a collection of rocks is concerned. In a few visits to this collection, the student will gain a knowledge of rocks, which he could never acquire from the hand-specimens in a common collection—a knowledge absolutely necessary to all true progress in the science.

49. The collection of rocks which we designate as the Wayside Museum is easily visited. Leaving the village of Lower

Invercloy at the smithy, and passing up the hill along the old Lamlash road, we are upon it at once. It is simply a stone fence bounding the road on the east side, and reaching from the brow of the old sea-cliff to the top of the hill. The rocks of which it consists were gathered from the adjoining fields, across which they were borne from the mountains and hills to the north and west, during the long period when the island was wrapped in its icy sheet.

As we pass up from the level of the sea to the top of the terrace—a height of five-and-twenty feet—the geological and picturesque features of the mountain nucleus come grandly out. In front the granite peak of Cior-Mhor, and serrated granite ridges of the Ceims (Kyims), Ben Talshan and Ben-Ghnuis, tower above the flat-topped, steep-sided foreground of slate and Old Red. On the extreme right, Maoldon, with its rounded top of fossiliferous sandstone, rivals in height the slate plateau S.W. of Goatfell. Planed off across the line of dip, and in the line of strike, these sandstone beds, with their intercalated limestones, stretch along the uniform line far down to Merkland Point, while their gentle southward inclinations form the charming woods and glades about the castle. Subsiding beneath the alluvial plain of Brodick, whose bay, excavated in their yielding strata, once reached as far as Glen Rosa, these sandstones, bearing the limestones with them, have been heaved up into a long narrow ridge, which culminates on the extreme left in the rugged porphyry of Windmill-hill, and rounded top of Ploverfield, whose granites and syenites reach thence across the moors to points out of sight, by the heads of the Bhein Leister and Clachan Glens. As we pass to the summit level of the road, the ridge of the Sheans comes in sight—knolls of common greenstone capping sandstone which isolates them.

50. As the position of the rocks is wholly fortuitous, we shall notice them in the order of their age. Granite occurs in several varieties, exhibiting distinctly the three ingredients, quartz, felspar, and mica. The felspar is either white or red,

more usually white in Arran, being always in such quantity as to give its colour to the rock, and is easily known by its laminated structure and rhombic form, which presents, in a fractured specimen, almost a square face. The quartz is harder, gray or vitreous, and, if in crystals, the form is a six-sided prism terminated by a six-sided pyramid. In our Wayside Museum no specimens are met with containing the smoke quartz or cairngorm crystals, which are found on the back of Suithi-Fergus and the Castles, and are of the same form as the common rock-crystal, or "Arran diamond." The mica is dark gray, dark brown, or almost black, in thin, flat, four-sided prisms, bevelled at each end with two surfaces, so that when these are equal in length to the others, the crystal becomes a thin six-sided prism. When hornblende takes the place of mica, and the rock consists of quartz felspar and hornblende, it is called syenite. Hornblende is easily known from mica; the crystal is a long narrow prism, not divisible like mica into thin flakes, and the colour is darker. The rock has a mottled, dark green, red, or white aspect, according to the colour of the felspar, and is much tougher than common granite. We have not met with it among the northern mountains, and it may have found its way here from the Ploverfield district. Some syenites contain also a little mica; but such structure has not been noticed by us in Arran. Another variety found here consists of quartz and felspar, and has either no mica or extremely little; it is called Eurite, and is usually tougher than other kinds of granite. It is the Weiss-stein of the Germans, and the Elvan of Cornish miners.

We do not find gneiss in our Wayside Collection—it does not exist in Arran, and is only found *in situ* far to the N.W. Mica slate, consisting of quartz and mica, which occurs on the N.W. coast, and in Cantire, is met with, as are the common slates enveloping the granite nucleus, among which the most remarkable is a very hard and tough banded slate, such as occurs *in situ* near the line of junction of the granite and slate, in the various localities already indicated. The bands are no doubt due to a re-arrangement of the colouring matter, probably iron, caused by the intense heat along the line of contact. Both parts are equally flinty. Masses of slate also occur composed of quartz and chlorite, chlorite taking the place of mica. This chlorite is of a green colour, and occurs in small very thin flakes, never reaching the size of plates, as mica does. There is very little of this chlorite slate in Arran; but it occurs in North Bute, alternating with mica slate, chiefly towards its outer border. A few masses of another slate occur, which consists chiefly of hornblende, but dashed occasionally with spots of felspar; it is almost like a basalt in structure, but coarser in grain, and more crystalline. It is extremely hard and tough. This is hornblende rock; it probably forms beds in the slate on the north-west coast, but we know no locality for it. Masses of red sandstone and conglomerate are frequent. The chocolate-coloured Old Red, such as is well seen towards the entrance of Glen Sannox, occurs, and another singular variety, a passage apparently from slate to sandstone. A mass of this will resemble porphyry in one part, in another it will contain pieces of imbedded slate, and again there will be finer layers like a fine semi-crystalline conglomerate or coarse slate. Large lumps of quartz are abundant, of several varieties; some derived from the conglomerates, and others directly from the northern mountains.

51. By far the most interesting series of specimens to be found here is that embracing the igneous rocks of later origin. Among them a regular gradation may be traced from coarse crystalline greenstone through basalt to pitchstone. These rocks, with felstone, affect a stair-like arrangement of beds when forming the surface of a district, sheet succeeding sheet upon the ascent of a hill, in a terraced form, like the steps of a stair. They are hence called Trap rocks, from the Swedish *trappa*, a stair. They are also called Plutonic rocks, as formed in the "nether depths," to distinguish them from the volcanic rocks, erupted at the surface. Felspar is the base of all these rocks; to this hornblende being added, the compound is

greenstone or basalt, according to the structure; augite combined with the felspar constitutes most of the modern volcanic rocks or lavas; hornblende belongs to the ancient igneous group; augite chiefly to the modern. The felspathic rocks, those containing felspar alone, are, viewed chemically, silicates of alumina, mixed with silicates of potash or soda, lime, iron, and manganese; when hornblende or augite are in combination with the felspar, they are silicates of magnesia, combined with the other silicates as above. And it is important to bear in mind, when studying these rocks in the field, that however different they may appear, they are all very similar in composition, the differences among them being due to different rates of cooling from a state of igneous fusion. Rapid cooling produces a glassy substance like the obsidian or volcanic glass of modern volcanic districts; but if the fused matter crystallise, when cooled to a certain temperature, a complete physical change takes place, and a stony structure is assumed.

The felstones of Arran are, felspar porphyry, claystone, and clinkstone. When a rock consists of any simple base with crystals of felspar imbedded, it is called a porphyry, and is named after the nature of the base. Thus felspar crystals, being imbedded in a uniform felspar base, the rock is felspar porphyry; so also we have pitchstone porphyry, hornstone porphyry, and so on. But the term is often used loosely, as when the base is not simple, and the imbedded crystals not felspar. Compact felspar, claystone, and clinkstone are mere varieties of one substance—a homogeneous felspathic rock, with occasionally bits of imbedded quartz, and consisting of about 75 per cent. of silica, 15 of alumina, and 10 of potash, soda, magnesia, and oxides of iron and manganese, with a trace of lime. All these felspathic traps or felstones abound in Arran.

52. At the head of the hornblendic traps stands greenstone. It is composed of felspar and hornblende, and varies in texture from a coarse crystalline rock with large distinct crystals of the

two minerals, to a fine-grained, compact rock, in which the component parts cannot be distinguished. The rock then passes into an almost homogeneous substance, of a black or dark blue colour, in which, by means of a lens, one can only distinguish minute shining flakes or points. This is basalt; and it is known to consist of the two minerals above named, mainly from the gradual passage which is observed from greenstone into it. Such a suite of specimens will be easily formed from the collection of rocks we are now speaking of. The chemical composition of the two rocks is very much alike, ranging from 45 to 60 per cent. of silica, 10 to 20 per cent. of alumina, 10 to 15, and sometimes 25, per cent. of oxide of iron, with small quantities of potash, soda, lime, and magnesia. It is chiefly the black basalt of Staffa and the Giant's Causeway that contains so large a proportion of iron. Other varieties of the hornblendic traps found here are those improperly called trap-porphyries. They are fine greenstones, containing imbedded crystals of felspar; and the base being thus compound, they are not properly porphyries. One is speckled with felspar dots; the other more homogeneous and dark-based, with distinct felspar crystals.

Greenstone is also found here in spheroidal masses, which are very interesting, as illustrating the origin of these rocks. The spheroids decompose in concentric coats; and are found to be finer-grained inwards, so that the nucleus is often a fine basalt. As the crust thickened by cooling, the heat would part from the centre more and more slowly, and the most perfect structure would be developed at the centre. In the great ash bed, or great ochre bed, at the Giant's Causeway, which is 25 feet thick, we noticed, some years ago, a similar structure. Rows of spheroids occur in this bed in some places, and in breaking them we found the external coats to be the common ash, or ochre, and the nucleus a fine greenstone, the intermediate coats presenting a regular gradation between these two extreme states. This structure is indeed singularly characteristic of the trap rocks, the slightest development of it enabling us at once to recognise them amid metamorphic slates; the columnar and jointed forms which these rocks so often put on being also dependent upon it. These relations have been made known to us by Mr Gregory Watt's series of experiments, instituted at the Soho Foundry, near Birmingham (*Phil. Trans.*, 1804). He fused the basalt of the Rowley-hills, and allowed it to cool at different rates. Small portions taken out and quickly cooled became a glass, like obsidian; others, allowed to cool more slowly, took the aspect of pitchstone; and when the heat was withdrawn from the furnace with extreme slowness, a perfect jointed basalt was the result. At a certain stage in the cooling of the semi-fluid mass, minute globules, enlarging into spheroids, were formed within its substance. From the centres of these there radiated distinct fibres, which divided at equal distances from the centre, so as to detach portions of the spheroid in concentric coats. When the radii of two spheroids touched at their extremities, the one set of fibres did not penetrate the other, but the two bodies became mutually compressed, and separated by a well-defined plane. When several spheroids came in contact, they formed one another, by their mutual pressure, into prisms with perfect angles. Each joint of a basaltic pillar is thus a compressed spheroid. The cup and ball articulation was perfectly formed in these experiments, and obviously arose from pressure, varying in amount and in direction. In proportion as the centre of radiation of the fibres became more remote, the articulations between the joints would approximate to planes. Actual dissection of a basaltic pillar by the hammer confirms this view of its structure. A great many small pieces may be detached all round a joint, leaving a spheroidal nucleus occupying its greater part; and in this a radiation from a centre may be seen. All our basalts, greenstones, and felstones exhibit this columnar and articulated structure; but the concave and convex jointing is seen only in basaltic pillars. If the centres were all symmetrically disposed, and the globules or balls first formed all of equal size, the columns would all be hexagons, since six would touch one all round. The varying forms of pillars, and their inclined, curved, or, as in the case of Dun-Dhu mentioned in our last Excursion, radiating or fan-shaped position, are undoubtedly due to an irregular distribution of the centres of radiation through the fluid mass. It is plain that dikes are *prismatic across*, because the heat passed off perpendicularly to the cooling surfaces.

Pitchstone, which figures so largely in the geology of Arran, is found in several varieties in our "Museum;" and when containing specks of white or red felspar forms pitchstone porphyry. Its composition is almost the same as that already given for felstone, being from 62 to 74 per cent. of silica, 11 to 17 of alumina, with small quantities of potash, soda, and lime, and oxides of iron and manganese. Its peculiar character is derived from its rapid cooling. We have to mention, lastly, a highly interesting rock found here, but never met with by us *in situ* in Arran; only a few blocks of it are seen in the fence. It is intermediate between basalt and pitchstone, and links them together as products of fire generated at different stages of the cooling process. It is hard and tough, and almost homogeneous like basalt, but without its shining points, while it has the semi-vitreous aspect and colour of a dark-blue pitchstone. It would be extremely interesting to find this rock *in situ*, and to mark its relations, whether occurring alone, with basalt, with pitchstone, or whether all three are together. Veins of it must exist somewhere to the N. or N.W.

53. Our walk homeward will be varied by passing down from the top of the hill into Birk Glen, along the old road. Having crossed the burn, we come on the great pitchstone vein lying a little diagonally to the path. It is of a green colour, lamellar structure, and is in parts porphyritic. The breadth is about thirty feet, and range between west and south-west. Thin white films form on the exposed surfaces, and when constantly moist decompose into a tenacious clay, coating the surface. It extends into the bed of the burn, and is there of

the same width; but the contact with the sandstone is not well seen. So far as can be judged there is little alteration on the sandstone. On the lower side a dike of hard, blue, fine greenstone is in contact with the vein; but the portion touching the dike is shivery and crumbling. Passing upwards to the new Lamlash road, indications of the vein are seen on the north of a claystone bed, which is quarried on the west side of the road. On the east side, some way above the road, the bed of claystone, formed below of rhombic blocks, but slaty above, is worked extensively for road metal; and on the north side of this quarry the pitchstone vein is seen coming against the claystone, and partly covered by it; a wing, as it were, of the claystone bed stretches partly over the pitchstone, so that it does not reach the surface; a curious relation rarely to be seen among the Arran dikes. It is quite possible, therefore, that it may be covered wholly or in part by the claystone bed or vein below the road, since it is exactly in its line of bearing. This relation of the two beds lends countenance to the idea that these claystones are but altered sandstones. It is remarkable that there is no trace whatever in the river bed of this claystone vein. The pitchstone is traceable all along the hill-side towards the N.E. to the crest of the moor; whence it bends round nearly S.E., and reaches the base of Dun-Dhu as already pointed out. This course implies undulations somewhat greater than any we have yet seen in dikes, some of them being fully 30°; but on the Corrie shore they reach 80°; and it is therefore quite conceivable, since the continuity is made out so far, that our pitchstone vein may bend round again after leaving Birk Glen, and re-appear at Brodick schoolhouse. Still it would be satisfactory to find it at some intermediate point.

This pretty Birk Glen would be fit for cultivation, but it has been long kept in its present state of wildness as an excellent cover for woodcock. We may return from it by the new road; but there is more variety in descending the stream and passing through the ornamental grounds connected with the hotel. The sandstone strata are well exposed, and there are some curious dikes. The burn throughout its whole course is highly picturesque; indeed none of the Arran burns, lovely as most of them are, have the same variety of picturesque features as those in the red sandstone. The alternation of hard and soft strata, of sandstones and conglomerates with variegated marls, the depth and peculiar style of disintegration of the rocks, the many pretty ledgy falls, the contrasts of the black basaltic veins, of the rich colours of the rocks, the luxuriance of the wild wood and wild flowers, bring out a variety in the effects not to be seen in burns running over granite, slate, or basalt. This is strikingly seen in the burns of North and South Sannox glens, Glen Scorodale and Torlin, which gain greatly in picturesque beauty directly on their entrance among the sandstone strata. In drawing and painting them, rocks ought always to be individualized; and every Artist ought to be a Geologist. A thorough acquaintance with the rocks would be like the implanting of a new sense.

EXCURSION V.

BY THE CORRIE SHORE.

54. The hotel at Invercloy occupies a charming situation on a terrace of the old sea-beach. The site was chosen by the late Duke of Hamilton, who had a true and exquisite taste in art and nature, and the keenest relish for their beauties. Perhaps the finest view is had on turning the north-east angle of the building, where the granite peaks and ridges, the wooded hill-sides, the castle, the sweep of the bay, and the opening of Glen Rosa are all in sight. When the mountains are clear of mist, the effects are always wonderful. They are perfectly magical sometimes in the early summer morning, when a gentle air comes into the bay from the cooler water outside, and masses of white vapour cover the middle region of the hills, the peaks projecting,

> Where many a pinnacle with shifting glance,
> Through the grey mist thrusts up its shattered lance,

and the lower edge of the vaporous mantle is like a veil of gauze, half concealing the castle and its woods. Slowly the veil rises, the vapour rolls upwards, now covering and now revealing the highest peaks, till, as the bare rocks get heated, the mist gradually melts away from the lesser heights, and all that now remains is gathered on the sheltered side of Goatfell, and raised by the eddies from the glens and corries into a cloudy pillar resting on the summit. Presently dissolved below, the pillar in its upper part mingles with the fleecy clouds aloft, and the peaks stand out all day long in unclouded grandeur.

From the hotel at Invercloy to the village of Corrie is a distance of six miles, but the sweep of the road round the bay takes up nearly two of these. A boat may be taken to the Brodick pier opposite, the Rosa burn may be crossed by a rustic bridge, but the sands will prove not a short route, and the river is usually impassable. We shall pass round by the public road. The Cloy burn formerly entered the sea at the lower village, as the name Invercloy implies; that course was changed by a bar of sand cast up during a storm, and the stream has now for a long time followed the winding course which unites it with the Rosa burn before reaching the sea. The ridge of gravel and sand dividing the beach from the marsh behind, is the joint work of the sea and the two streams. This marsh, and parts of the gravel ridge which nourish coarse brushwood, are the favourite resort of many plants—as the Isle of Man cabbage, the sea purslane, the sea pink, catchfly or sea campion, the scurvy grass, arrow grass, two plantains, and many others.

Brodick church, on the platform finely overlooking the plain, was formerly a chapel of ease to Lamlash church; but Brodick is now erected into a separate parish, in so far at least as *quoad sacra* purposes are concerned. For secular purposes, the old division into two parishes still subsists. That division was made in a singular way: the western half of the island, from Loch Ranza to Largiebeg, formed Kilmorie parish, the eastern half Kilbride parish, the boundary line running along the crest of the dividing ridge of the island. The north part of Kilbride parish forms the new parish of Brodick, which extends northwards from the top of the ridge between the bays of Lamlash and Brodick. The first minister of this new parish was ordained on the 7th April in the present year. An endowment has been provided, jointly by the late Duke of Hamilton and the Endowment Committee of the Established Church.

Passing eastwards from the Rosa bridge and then north towards Brodick pier, and emerging from the belt of planting, we are on the site of the old village of Brodick, and obtain one of the sweetest views in the island. The lower reach of Glen Rosa, with its grand background of mountains, is seen across the rich glades of the park, covered with pasturing sheep, and finely varied with tree clumps, and tall standing stones. The enlargement of the park, the formation of a new approach, and the opening up of this fine view, rendered it necessary to demolish the village. Yet had the old place a beauty of its own. The situation was delightful, and it was pleasant to see, on a fine summer evening as one returned from a walk among the hills, the happy village groups enjoying themselves out of doors; the young sporting on the common, or strolling in the green lanes; the matrons bustling to and fro; the men in knots about the cottage doors. But the household gods have sought a new sanctuary at Upper Invercloy, and the population is now out of sight.

55. Between the fine picturesque group of ash trees near the beach and the Brodick pier, the strata of sandstone rise from beneath the sands, and exhibit some extraordinary irregularities of stratification. Two sets of beds dip out of the usual direction and at higher angles. The usual dip is due S. by the compass, that is S. 26° E.; the irregular beds make angles with these. Those on the E., next the regular strata, dip E. 34° S. The other set west of these dip due E.; and when in the next set the usual south dip is resumed it is 10° farther round toward the E., that is S. 36° E., instead of 26°. There is no apparent cause, in dike or fracture, for this strange tumble-about among the beds—it seems more probably due to eddies at the time of deposition.

East of the pier, the strata are also slightly irregular, and the angle of inclination greater than usual. The beds near the pier dip S. 46° E., and a little farther E. the dip is S. 21° E.; the former being a difference of 20° from the dip W. of the pier. We shall find similar examples of these irregularities farther along the shore. By the roadside here, under the summer-house perched aloft upon the cliff, there is a huge granite boulder, the weight of which cannot be less than from six to eight tons. Somewhere about here, or near the old hotel, past which is the most approved path to Goatfell, there formerly projected over the cliff a very large ancient yew-tree, called the "hanging yew;" hanging, alas! in a double sense; for upon it, in the wild rule of some of the old bailiffs or stewards, who in the absence of their lords held despotic sway in Arran, summary justice, without trial, was wont to be executed.—But we must pass up to examine the castle.

56. Brodick Castle stands on a terraced platform of the sloping hill-side, at an elevation of 125 feet above the sea level, towards which several successive terraces descend to a distance of 400 yards, ending in the old sea cliff, here about 25 feet in height. The situation is thus fine and commanding, and the views from the principal apartments extensive and beautiful. The building is in the old baronial style, with battlemented roof, and is three storeys in height. A lofty tower, with terraced gables, and flanked with turrets, forms the south-west corner of the building. From the battlements of this tower the sea and mountain view is very grand. The alterations and additions made by the late Duke, when Marquis of Douglas, and which were completed in 1844, have fully doubled the capacity of the building, and harmonise admirably with the ancient portion. The entrance is by the west front; and the building terminates eastward in a ruined tower, clad with ivy and stunted fuchsias, whose erection dates back at least 300 years, and which is carefully kept in preservation. The handsome grounds about the castle are laid out in a style admirably suited to the situation. Off the south-east angle of the castle there is a beautiful flower garden, sloping southwards from the base of the first terrace, whose summit is crowned with a row of splendid old plane trees; and there are many fine limes, beeches, and other trees, in the older portion of the woods about the castle. The disposition of the woods towards the north has been greatly improved within a few years by transplanting.

A stronghold of some kind seems to have existed from very early times on or about this spot. It was a place of great strength and considerable size in the time of Bruce, who, after his descent upon Arran, from Rathlin, besieged and took the castle, then held under Edward I. by Sir John Hastings; and from Brodick he is said to have taken his departure for the mainland, on his perilous expedition for the liberation of Scotland. This was in 1307, and there were still seven years before him of peril and varied fortune ere his authority was finally established at Bannockburn, on 24th June, 1314. The history of the various occupants of Brodick Castle will be found detailed with considerable fulness in Mr M'Arthur's History of Arran, and Mr Reid's History of the County of Bute.* The castle was razed to its foundations in 1544, by the Earl of Lennox, sent by Henry VIII. on an expedition against Scotland, so that no part of the present building can be older than that date.

The connection of the Hamilton family with Arran dates from 1474. Sir James Hamilton, created Lord Hamilton, his manor house of *Orchard*, in the barony of Cadzow, being declared his principal messuage, and having its name changed to *Hamilton*, was attached to the Princess Mary, daughter of James II., and would have espoused her but for the untimely death of her father. Her hand was afterwards bestowed by her brother, James III., on Thomas, son of Lord Regent Boyd, and, as her dowry, a grant of lands in Arran was made, and Boyd was created Earl of Arran. That family, however, soon fell—their estates were forfeited, the Earl fled abroad, and died there. In 1474, Lord Hamilton, now a widower, married his first love, the widowed Princess Mary, but survived only five years, dying in 1479, and leaving a son, James, second Lord Hamilton, then only four years old. By charter, dated 10th October, 1482, the forfeited lands of Regent Boyd and his son, the Princess' first husband, were made over to her in liferent; and in 1503, her son, now grown to man's estate and having already rendered important services to the Crown, received (August 11) a charter of the lands in Arran, and the

* Both works are published by Messrs Murray & Son, Glasgow.

title of Earl of Arran; and by charter, dated the next day, a gift "of all the castles and fortalices, mills and fishings, with patronage of all churches, and a commission of justiciary."* The honours and properties of the family steadily increased from this time, fortune never for a long time forsaking them. They never lost hold of Arran, the whole island belonging to them, except the estate of the Fullarton family. For about fifty years, Lord Rossmore held all the north-west portion of the island, from Iorsa waterfoot to Loch Ranza, except Whitefarland, part of the Fullarton property; having received it, being unentailed, as a marriage dowry with Anne Douglas, a daughter of Douglas, eighth Duke. But Lady Rossmore dying without issue in 1844, the Hamilton family soon after acquired the property. Anne, eldest daughter of James, first Duke, Duchess in her own right as heir of her uncle William, second Duke, who fell at the battle of Worcester, fighting on the royal side, did much to improve Arran, and her benefactions are still remembered by the islanders. She died in 1716, at the age of 80. But perhaps no member of this family has done so much towards the physical and moral improvement of the people as the late Duke, William Alexander Anthony Archibald. He greatly improved the means of communication between the different parts of the island, encouraged and aided the farmers in draining not only the arable land, but also the hill pastures; built new hotels, built and endowed schools, and endowed a new church. The improvement of the castle and park are also due to his taste, and he built the fine lodge on Dippen cliffs. His Grace's munificent patronage of art, and exquisite taste are well known. His Grace, when yet in middle life, died at Paris, July 15, 1863. His memory will long be cherished, as is that of the Duchess Anne, by the grateful islanders.

57. Between Brodick pier and the gate of the principal approach to the castle, there are many curious dikes and

* Reid's History of the County of Bute, p. 176, *et seq.*

irregularities in the strata; indeed, the whole coast from this point to Corrie is worthy of careful study, as illustrating the natural history of dikes; and will prepare the geologist to expect a broad dike to become a thread, and not to be surprised if he see a dike "doubling" at an angle of 90° and "throwing" him completely "out." Between the granite boulder and the gate there is one dike which is, as usual, prismatic across, and breaking in westwards among the strata becomes a bed parallel to them, and presents on its upper surface a miniature "causeway" or pavement of prisms of various forms, placed perpendicularly to the surfaces of the bed. The case is very interesting, and is a good illustration of the remarks made in Art. 52. A little way east of this there are some other dikes on the flat beach which change their directions in the most sudden and puzzling way; and under the wall opposite the gate a singular contrariety of dip in the beds; strata dipping W. come "end on" against those dipping S. The dike here which ranges due N. and S. by compass, and the others already mentioned, may have something to do with this irregularity. Near this point is the landing-place called the Wine-port, where in the jolly old times, when the retainers were feasted in the castle, wine and beer were brought ashore in hogsheads; and farther up near the cliff are shown the remains of an old burying-place, where some important personages are said to have been interred. But of this there is no special record.

It would require a separate memoir on this coast to detail all the curious particulars that might be mentioned. We can only state in general that in the bay here are some broad dikes with remarkable bends, greatly altering the sandstone, and from their width and direction clearly identifiable on the Strathwillan shore opposite; that several singular inversions of dip occur in the sandstone beds, whose surfaces also are worn in a singularly uneven manner, so that portions stand out high above the others; that about Carlo cottage and the smithy, and between this point and the sandstone quarry, near

Corrie, there will be seen several dikes which bifurcate, and in the angle of separation enclose altered masses of sandstone; and that equally striking cases will be met with of a change in direction among these dikes. After examining them, no one will hesitate to consider it quite possible that the pitchstone vein of Birk Glen, traced all the way across the moor from the base of Dun-Dhu, may be the same which shews again at Brodick school-house.

The old sea cliff is worn everywhere into caves by the long-continued action of the waves. One of the largest of these is "Lily's cave," west from Carlo cottage, named from a poor woman of this name, who took it for her permanent abode, when her cottage was taken down to make way for some former improvement, the date of which is uncertain. She was a widow, with one daughter. The youthful and pretty Lily had a lover in good circumstances, whose truth stood the test of misfortune; but though he had good hopes of Lily's favour before, she would not listen to his proposals now—the widow would not leave her retreat, and Lily would not desert her. But soon the fair flower drooped and withered; and before a year had passed, the widow was left alone in her new dwelling. A gang of gipsies occupied the cave after the poor widow, and became a trouble to the neighbourhood. They were expelled by force; and, to prevent any new occupancy, the gravel bank above the rock was digged down and the mouth obstructed.

It is rarely that shells can now be found in any of the caves or hollows of this cliff. The rapid growth of vegetation, under the shelter, sunshine and moisture, has covered the floors with a deep deposit, in which it is hopeless to search. We have added nothing to the list originally given by Mr Ramsay, and which comprises the following:—

Cardium laevigatum.	Turbo rudis.
Patella vulgaris.	Trochus cinerarius.
Lucina radula.	——— magus.
Purpura lapillus.	——— crassus.
Turbo littoreus.	Nerita littoralis.

Venerirupis decussata.	Cytherea exoleta.
——————— palaestra.	Terebra reticulata.
Mytilus edulis.	Rissoa calathisca.
Buccinum undatum.	—— semicostata.
Venus fasciata.	

Three of these—the venerirupis decussata, rissoa calathisca, and trochus crassus—are now rare in the West of Scotland. These species therefore seem to connect the beds in question with those containing species now extinct in the seas around Britain.—The young geologist needs to be cautioned about drawing any conclusion from a few scattered shells found on the hills, as birds often carry shells even of considerable size up among the mountains. It is only a regularly formed bed that affords evidence of physical changes in the surface.

58. As we approach the village of Corrie, the sea cliff becomes lower, and on the edge of it there rest two huge granite boulders, one on each side the White water, which here comes tumbling down from the corrie on the north-east side of Goatfell. They are true boulders, granite resting on sandstone. The larger of the two, and the largest boulder in the island, lies on the north side of the stream. We estimate the weight at more than 2000 tons! What force could have hurled a mass so enormous from the mountain-side? Not gravity urging its descent; unless, perhaps, the first impulse was given by an earthquake, for the slope of the hill-side is gradual, and the mass itself ill adapted for easy rolling. But if we admit the agency of ice, we have a force adequate to the transport. This transport took place before the present level of land and water was established; the cliff was not yet cut out from the sloping hill-side; the bed on which the boulder now reposes was the sea bottom, and in the glacial period it may have glided down upon icy sheets descending from the corries, or been borne off by a berg which deposited here the heaviest portion of its load. A sister block rests upon the same cliff, on the south side of the burn; but its dimensions are much less. The shore is strewed with multitudes of similar masses of all sizes.

From this point the geologist ought to pass up to examine the limestone beds, thence to the top of Maoldon, and so across to the granite junction at the fall of the White water. The limestone bands come out in the broken ground in front of the north cliffs of Maoldon, dipping S.E. with the sandstone at angles of 30° to 40°. They are similar to the beds at Corrie, with the same number of integrant strata, each of which, however, is thinner than at Corrie, and with the same assemblage of fossils. They have been already noticed (Art. 13) as probably an upthrow of the Corrie beds; those in the woods N.W. of Brodick Castle may have the same origin; there being thus, in all, four upthrows of this stratum, separated by four great faults. Our reasons for doubting this view have been stated already (Art. 13).

If such a dislocation as is supposed really took place, there ought to be evidences of it on the shore, where all the strata are well exposed. Certainly none of the limestone beds of Maoldon are seen upon the shore. We cannot agree with the view stated by Murchison and Sedgwick, that there is a series of advances of granite from the central ridge. From the north side of Glen Sannox to the mill-dam, the line of junction of the granite and slate is extremely uniform, unbroken by spurs advancing from the granite nucleus. We believe that when the paper on Arran by these distinguished geologists was written; and even six years later, when Mr Ramsay wrote, the fact was not properly recognised, that in the Scottish carboniferous system bands of limestone may occur anywhere —from top to bottom of the series.

There is little else worthy of notice here; it is a wild bosky place, encumbered with huge fallen masses of sandstone, and with granite blocks, among which many pretty ferns and flowering plants find suitable habitats. But there are adders also about, and it is well to be watchful. The N. and N.E. fronts of Maoldon are very precipitous; the western part of the precipice is intersected by a greenstone dike about fifteen feet broad, ranging N. 25° W., and inclined towards the E. at an angle of about 15°. It forms a deep chasm in the cliff, over which the sandstone rises in a lofty wall. The amount of wearing here is prodigious, and we cannot conceive how it can have been effected without the action of the sea. Yet the place is many hundred feet higher than the ancient level, indicated by the raised beach so often mentioned already. We must, therefore, call in the agency of the sea, during the elevation of the land at an earlier period. Through a narrow passage between the dike and the sandstone wall, we can ascend to the top of the cliffs. A remarkable alteration is produced in the sandstone by this great dike. The top of Maoldon, about 1100 feet high, is composed of soft red sandstone, dipping S. 25° E. at angles of 15° to 20°; and thus the strata are "end on" to the granite. Impressions of plants may be found in the sandstone. The summit is strewn with granite blocks of moderate size.

In crossing from Maoldon to the fall of the White water, we pass the old red sandstone and slate bands; but no junctions are visible till we reach the base of the fall where the burn issues from the wild corrie. Judging from the position of the granite masses on the hill-side, one would not expect the junction so far up—it takes place at the base of the great fall. The best way to approach it, and to observe the changes in the strata, is to pass up by the south side of the stream. Here the usual chocolate-coloured Old Red is seen, and in it a great bed, apparently not a dike, of decomposing greenstone, with veins of calcspar and arragonite. Gradually the sandstone becomes harder and more slaty, and at length almost a quartzite. This is followed by a blue, very hard, flinty slate or quartzite, streaked with white, and in some parts containing felspar crystals, succeeded in its turn by a nearly white quartz slate, similar to that already described (Art. 44). It has in many parts a blue tinge, as if from transfusion of the colouring matter of the slate. This quartzose slate is several yards thick, and is in contact with the granite, here a hard and tough, close-grained compound of quartz and felspar, that is, eurite; the mica, if occasionally present, being in very small quantity. On the south bank, over the junction, there is a mass of the ordinary dark slate; it shows twisting of the layers and banding with blue and white, but is less altered than that in contact with the grey quartzite in the stream. The metamorphism at this junction is very remarkable, but the student must be prepared to encounter many difficulties in his ascent of the channel; and to submit, now and again, with the best grace he may, to a shower-bath from the dashing spray, as the water bounds from ledge to ledge of the long fall.

One can easily pass from this point to the top of Goatfell in three quarters of an hour; but we must now turn our steps to the Inn at Corrie.

EXCURSION VI.

To the North Shore.

59. The coast section from Corrie to Loch Ranza is full of interest. A long summer's day will be required to examine it carefully—several days, if we stop to collect the fossils which abound in some of the beds. The scenery, too, is full of beauty, in parts bold and picturesque in the extreme: we shall be tempted to stop very often to add some new treasure to our portfolio. Leaving Invercloy by the first steamer for Glasgow, we shall be at Corrie for an early breakfast, and be ready to start fresh while the day is yet young. But Corrie is, for this, as for several other excursions, an admirable point of departure. A handsome and commodious hotel has recently been erected here, and under its present management is most comfortable. Loch Ranza is eight miles distant from Corrie by the high road; by the path we are to follow along the shore, not more than between ten and eleven.*

The strata at the base of the carboniferous system, and their contact with the old red sandstone, are well exposed upon the shore. We shall trace them in ascending order, beginning a quarter of a mile north of Corrie, at the march of Achab farm, where the road bends toward the N.W. This is the base of the series; but it is not a well-defined base—there is, in fact, a gradual passage from the Old Red system into the carboniferous strata. The Old Red is here a conglomerate; and is overlaid by a limestone with imbedded pebbles, the same as those in the conglomerate, forming a calcareous conglomerate. This bed is followed by gray sandstone and concretionary

* Mr Douglas, postmaster at Corrie, and his son, are highly intelligent and obliging persons, well acquainted with the strata in the vicinity, and with the geology of the island generally.

limestone, or cornstone, consisting of red nodules, imbedded in shale. Other alternations of limestone and sandstone follow, there being in all five or six beds of limestone, the two lowest of which are seen rising up the hill on the dip S.E., over the conglomerate. The whole series is to be regarded as perhaps rather the bottom beds of the carboniferous series than the top beds of the old red sandstone. Then succeed these various beds of sandstone and shale, till we reach an enormous vein or dike of trap, which occupies the shore for more than 300 yards. In this are found basalt, greenstone, amygdaloid, and concretionary trap, exfoliating in concentric coats; and it is traversed by numerous veins of calcspar, steatite, and quartz. The next beds seen are a red and a gray limestone; there is then a whin dike, south of which, from the gate upon the road to the limestone quarry, the shore is occupied by sandstone. The strata of limestone are seen in the little harbour and extend up the hill, rising towards the north, and dipping S.E. at 36°. The rock is of a bluish-gray colour, about twenty feet thick, and consists of twenty-two beds of limestone, interstratified with the same number of beds of red shale, the thickest stratum seldom reaching one foot. The workings are inclined adits descending in the line of dip. Over the limestone are shales with hematitic iron ore. Several dikes traverse the limestone, and alter its structure; fossils abound, but are procured with difficulty. It has been remarked, as indicating the tranquil nature of the deposit, that the large producti uniformly rest with the convex side of the valve downwards. In the sandstone upon the shore in front of the village several common species of coal plants are found. In the little bay under the hotel there is a bed of limestone, and another a little south of Corrie, near the fall on Lochrim burn and the freestone quarry, that under Maoldon has been noticed already (Art. 58). The many interesting trap dikes have also been noticed; and those curious irregular ridges on the surface of the sandstone mentioned as occurring on the Corriegills shore, are even more remarkable here. They sometimes stand up six or eight feet above the surface, and have very sharp, jagged edges. In most cases they are independent of whin dikes, and consist of matter originally less liable to disintegrate than the sandstone, and probably introduced into fissures in this rock after it had consolidated.

60. We return now to the contact of the carboniferous rocks with the old red sandstone at the march of Achab farm, and follow the latter rock northward along the coast. Expanding inland, it rises into high cliffs. The forms into which these have been moulded by the action of the sea, the disposition of the natural wood upon them, and the huge granite blocks which stand prominent by the road-side, give a unique and most picturesque character to this part of the coast. One of the largest blocks (on the western edge of the road among trees) we estimated at above 200 tons; farther north on the eastern edge of the road is another very large block standing upright on its apex, perhaps let down so from a floating berg, or originally imbedded in sandstone, afterwards worn away as the tide ebbed and flowed around. The force exerted by a strong man causes it to oscillate slightly as a rocking stone. A melancholy interest attaches to another boulder south of the march of Achab farm. A garrison of eighty men had been left in Brodick Castle by Cromwell; against these the natives became so irritated, on account of the excesses of some of the soldiers, that when they were out upon a foraging party in this direction, the Arran men rose against them, and put them all to the sword, except one poor fellow who escaped, and hid under this stone. His place of concealment being soon discovered, he shared the fate of his companions.

We cross the Sannox burn by a rustic bridge, where it comes flashing down along its bed of granite sand, among wild copsewood, through green shadows and gleams of sunlight. Our path is now by the farm-house on the shore; but we must not pass so near the finest view of Glen Sannox without stopping for a little to look upon it. This point will be readily found upon the road towards North Sannox house. We look down upon a broken foreground, sloping on both sides towards the stream. On the right the glen is bounded by the long, steep ridge of Suithi-Fergus and The Castles; on the left by that of Cioch-na-Oich and the prolonged ridge of Goatfell. Cior-Mhor stands proudly up, closing the long vista.—We return to the shore. The high ridge of ground dividing North and South Glen Sannox terminates on the shore in a precipice called the Blue Rock, from a decomposition taking place upon it, due to the presence of iron and manganese, acted on by trickling streamlets. Here, and along the shore, the strata of Old Red still retain their southern dip, at a small angle. But when we cross the North Sannox water, we find the inclination southwards much less; and as we advance a little, the strata become horizontal. Still advancing, we find them dipping in the opposite direction, at angles gradually increasing till the original dip of 15° or 20° is reached, but now directed N.N.W., instead of S.S.E. The line *from* which these opposite dips are thus directed, and which is nearly in the direction of North Sannox Glen, is called the Anticlinal axis. The name was applied by Murchison and Sedgwick; but the relations of the strata were first pointed out by MacCulloch. It is obvious, then, that the strata are successively newer as we advance along the shore towards Loch Ranza. We shall pass over their basset edges rising southwards, the dips being northerly, and the inclinations on the shore from 10° to 15° and 20°; while on the face of the hills above they are from 50° to 70°, but in both cases alike towards the same point of the compass.

The Old Red here is cut by several dikes, of which two are extremely interesting. One of these is about half-a-mile north of North Sannox burn; it entangles a mass of altered sandstone, wedge-shaped at either end, five yards long and seven inches thick; the range is nearly north-west. The other dike is at the angle of the shore, where the Fallen Rocks first come into view, at little more than half-a-mile distant. The dike is best seen in the sand under the grassy bank. The structure of the trap varies, much of it being a fine blue greenstone or basalt. The change on the sandstone is remarkable. The interlacing of the two rocks, and intrusion of string-like veins of trap among the sandstone strata, as forcibly attest as any dike in Arran the irruption of a liquid stream of lava into the crevices of a fissured mass.

The Fallen Rocks are about two miles from Sannox; they are an immense debacle of masses of old red sandstone hurled from a hill above, where an overhanging cliff gave way; and now strewing the beach and steep slope in magnificent confusion. They seem freshly fallen, yet Headrick described them fifty years ago as we see them now.

61. At Groggan point, immediately west of the Fallen Rocks, we come on the lower beds of the carboniferous series following conformably on the old red sandstone. They are very similar to the corresponding beds at the march of Achab farm, near Corrie; a calcareous conglomerate, with beds of white limestone, being the lowest. These are followed by sandstone and shale, with nodular limestone and thick bedded red and white sandstones, till we reach the limestones near the Salt Pans, which are the same as the Corrie beds. But we must not omit to notice the great trap beds of Lagantuin bay, which upturn the shales from their usual inclination of 20° to an angle of 58°. These traps are—amygdaloid with zeolites, steatite, and carbonate of lime, claystone, and porphyritic greenstone or trap porphyry with crystals of diallage; and it is remarkable that their position is the same as that of the great trap bed at Corrie. The coast section near the coal pits is very interesting; it is thus given by Murchison and Sedgwick, as seen west of Millstone point:—

1. Blue shale, with limestone, ... 25 feet.
2. White sandstone, ... 2,700 „
3. Carboniferous limestone and shale, alternating, 120 „
4. White sandstone and variegated shales, . 2,000 „

Sandstone and shale, containing seams of coal, succeed these till the beach is obscured by shingle. The principal seam is three or four feet thick; but has been worked out, as far, at least, as the level of the sea, which now invades the seam. It was used only at the adjoining salt and lime works, and never exported. There were workings on two or three seams in the direction of the dip, which is nearly at 45° north, but all were abandoned when, from the depth of the workings, the sea gained access to the pits. The coal is of that variety called *blind coal,* containing an unusual quantity of carbon, and burning without flame or smoke. Many vegetable impressions, chiefly ferns and calamites, were found in the coal below the shale; the stems of the latter being formed of a hard coal with a bright shining fracture. This character of the coal may be partly due to the action of several dikes which here cut the strata. An adit was opened in the hill above on the strike of the seam, but no coal was found. Still there seems little doubt that there must be considerable beds of coal yet untouched. The shale and coal tract is bounded by a black limestone below and red limestone above, 1,400 yards apart, and extending up against the schist precipices, thus affording ample room for a considerable development of the coal beds.

A little to the west of this, beyond a shingly beach, sandstones and shales again appear. The shales contain beds of ironstone, some of which have the structure of septaria; and these, with the variegated shales, form on the shore a flat platform, with a tesselated appearance, like a mosaic pavement. "I doubt," says Mr Headrick (p. 210), "if the most skilful mason, or even a mathematician, could produce anything more regular or more beautiful."

Farther on—about half-way between the Salt Pans and the "Cock of Arran"—several beds of red limestone, rich in fossils, and of red shale, occur. It is interesting to notice the perforations of pholades in the limestone above the level of the present tides, as being a striking collateral proof of that change of level to which we have so often alluded.

These limestones are succeeded by a series of sandstones, shales, and fine conglomerates, overlaid in their turn by variegated marls containing nodular ironstone, and by white sandstone. Over the latter, a little way east of the Cock, lie beds of fine red sand, alternating with fine conglomerate, the dip of both being 65° W. of N., at 23°. These beds, with the white and red sandstones which succeed them westwards till the schist is reached beyond the Scriden, are the uppermost members of the whole series, which begins to overlie the old red sandstone near the Fallen Rocks. From their position and mineral character they have been classed as new red sandstone by Murchison, Sedgwick, and Ramsay. Mineralogically, they have a much greater resemblance to lower permian strata than any of the rocks in the southern district; but we refer them, notwithstanding, to the upper carboniferous series, till fossils shall be found which may decide the question.

The Cock of Arran, near which this series begins, is a large isolated mass of sandstone, resting on the beach, a noted landmark among sailors. When seen in front from the sea the block had the form of a cock, with expanded wings, in the act of crowing. The resemblance is now less striking, as the head has been broken off. Beside this block there are two singular whin dikes close together, on the flat beach, about two feet wide. They terminate in the sandstone, almost opposite to one another, and are prolonged in contrary directions.

62. The Scriden is a headland whose base is strewed with immense masses of sandstone. These fell about 100 years ago with a loud noise heard in Bute and Argyle. The debacle is more extensive than that of the Fallen Rocks, but inferior in grandeur. It was produced by a landslip of the mountain side, the traces of which yet remain in a long deep rent near the summit. Scrambling for a long way among the fallen masses, we reach an open shingly beach, along which the line of the old slate, advancing from the interior, strikes the shore, and cuts off the red sandstone; but the nature of the ground does not permit the junction to be seen. The slate here dips about S.E. at 40°. The change takes place near a glen, with a burn called Alt-Mhor (large burn). The variety called chlorite slate occurs here, and quartz abounds in veins and beds in the slate.

A short distance forward, at Newton Point, where the coast bends round into Loch Ranza, and a small stream, called Alt-Beith (birch burn), enters the sea, there occurs one of the most instructive sections to be seen in Arran. Strata of sandstone again occupy the shore for 300 yards, dipping into the sea, and resting along the platform in front of the cliffs upon the upturned edges of the strata of slate. These make an angle of 40° with the horizon, and dip 40° E. of S.; the strata of sandstone are inclined at 25° and dip 55° W. of N.; the dips being thus nearly in opposite directions (see Art. 5, *sub fin*). This unconformability indicates that the slate, itself a sedimentary deposit, had not only been formed in this regular stratification, but had undergone a general disturbance, before the sandstone beds were thrown down upon it. The position of the slate-strata has no relation to the granite of the nucleus; the dip and strike are related to the great axis of elevation traversing the country in the direction of the Grampian chain. This position had been acquired before the Arran granite was injected amid the strata of slate, and ere yet any of the sand-

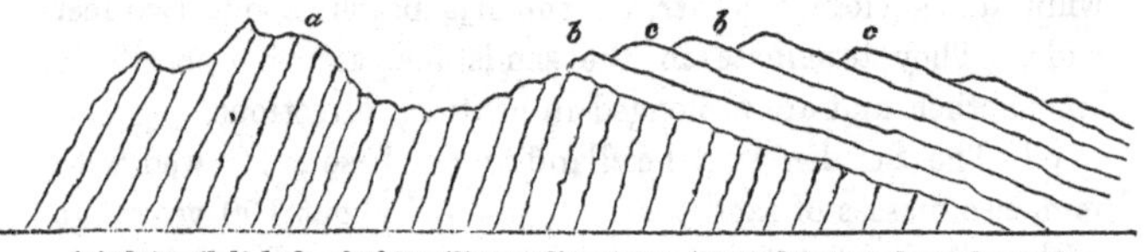

(a) slate; (b b) beds of white siliceous limestone; (c) sandstone and conglomerate.

stone beds which succeed it had been deposited. These sandstones (old red and carboniferous) are conformable to one another, and the deposits blend at both sides of the section, Achab farm and the Fallen Rocks. But we see here, as repeatedly noticed already, that neither deposit has any relation to the stratification of the slate, which had sustained extensive dislocation before the deposit of the old red sandstone had begun. The cut annexed represents the appearances at this place.

The sandstone strata here alternate with beds of limestone and conglomerate. The lowest bed, *b*, next the slate, *a*, is a hard crystalline white limestone, about six feet thick; it contains quartz pebbles, schist, and diffused siliceous matter, and is without fossils. There are several beds higher in the series, but the thickness is less. The total thickness of the various beds is not more than fifty feet. Whin dikes traverse both the slate and sandstone. The presence of this peculiar limestone fixes the age of the deposit, which resembles, in all respects, that already noticed at Achab farm and the Fallen Rocks, as forming a gradation from the Old Red to the carboniferous series—there being a difficulty in determining to which it ought to be assigned. We are inclined to the view that it is the lowest member of the carboniferous system. The occurrence of the limestone here has been noticed by Jamieson (*Min. of Scot. Isl.*, vol. i., p. 78, edit. 1800), from whom our cut is taken; by Headrick (p. 206, 1807); and by MacCulloch (vol. ii., pp. 376, 356, 1819). Later writers have classed it as new red sandstone, overlooking the limestone, and have given the dips erroneously. The deposit is a mere isolated patch, and has no connection with the sandstones eastwards, which abut upon the slate, and are clearly the uppermost members of the carboniferous series in this part of the island.

Very few granite blocks occur along the north shore, and none of more than two to three tons in weight. · But as we enter Loch Ranza and its glen, opening directly on the granite nucleus, we meet with boulders of great magnitude. Two of this kind, with many lesser, rest on the beach at the north-east angle of the bay.

63. One generally reaches this point late in the day, when the rays of the declining sun already fall strongly on the outer part of the bay, the upper portions being in the deep shadow of the high western hills. The old castle on its raised bank of shingle, which has kept its place steadfastly against wind and tide for at least 2,000 years, is a fine object in the foreground. The middle of the picture is occupied by boats, in from the fishing, and by nets suspended between high poles, beyond which is a line of white cottages on the west side, and houses of higher pretensions on the east, from which the curling

smoke rises up invitingly. The background is a dark circle of gloomy hills, now reflected in the lipping tide, which give an air of peculiar solemnity and seclusion

> "To the lone hamlet, which her inland bay
> And circling mountains sever from the world."

Beyond, and over all, rises the serrated granite ridge of Castle Abhael and the Sui, gleaming now in the golden light of sunset, and contrasting strongly with the dark slate hills of Ben-Leven and Tornidneon, thrown into deeper shade by the high hills on the west. The place altogether has a singularly picturesque and unique aspect; the stranger will say he has never seen such a hamlet and bay before. Yet will he find most comfortable quarters at the unpretending inn, and even luxuries not easily found elsewhere.

The stranger who has got but one night at his disposal to spend in Loch Ranza, should contrive to reach the place so early that he will have time for a stroll along the sea-shore, to the west, in the direction of Glen Catacol. Towards sunset, on a clear and still summer evening, the scenery appears to the highest advantage. It has then a quiet but touching beauty, which steals into the very soul.

There are several objects of much interest about Loch Ranza. The castle is a regular structure of stone and lime, with thick walls, and is still in tolerable preservation, though now roofless. There are several small apartments and a large hall reached by a narrow stair. It must have been a place of great strength before the use of artillery was known. It stands on a bank of shingle, running across the mouth of the bay, and forming the harbour. We first find it mentioned as being a hunting-seat of the Scottish kings, in 1380; it was then reckoned one of the royal castles. But the date at which it was erected is not known. In 1452 the lands and castle of Loch Ranza, with the lands of Sannox, were granted by the Crown in feu to Lord Montgomerie; Ronald M'Alister, who held them before, having refused to pay rent on a plea not deemed sufficient. The Montgomeries of Skelmorlie, ancestors of the Eglinton family, got a grant of the property in 1685, with much of the north end of the parish. They, however, lost the entire property, as an unredeemed mortgage for £3,600, to the Hamilton family in 1705. The Duchess Anne, then the head of that family, built a chapel and established a missionary station at Loch Ranza, for the maintenance of which she mortified a sum of money, producing £27 a-year. No trace of the chapel now remains. The ruins of the convent of St Bride, which till lately existed here, are also swept completely away. Some remains indicating the situation of an old cemetery have lately turned up near the supposed site. The only interest connected with it lies in association with the well-known scenes in Scott's *Lord of the Isles,* in which the scenery of "fair Loch Ranza" is very prettily and graphically described. The harbour is the resort of fleets of fishing boats. It is a most picturesque and exciting scene to witness the launching of these boats on a summer evening, under a favouring wind. In rapid succession they drop down from the harbour into the outer bay, and dart away in bounding glee, some out north into Loch Fyne, and others away to the left, across the broad waters of Kilbrannan Sound. Though so land-girt and safe-looking, Loch Ranza is by no means a safe anchorage. The most fearful squalls, sometimes even in summer, fall upon it from the narrow opening of Glen Eais-na-vearraid, between Tornidneon and Meal-Mhor. The currents of a S.E., S., or S.W. wind traverse the whole length of this long glen, hemmed in between high hills, and issue from the narrow gorge with terrible violence. Vessels are thus often driven from their moorings, and obliged to seek safety in the open frith, or by running for Loch Fyne.

EXCURSION VII.

BY THE WESTERN SHORE.

64. THREE interesting excursions may be made from Loch Ranza, from none of which, however, will it be easy to return to this place on the same day. They may, therefore, be taken at different times, after an interval. Most tourists will prefer the last—a return to Corrie by the inland route.

A walk by the shore, from Loch Ranza to Dugarry, about eleven miles, will more amply reward the lover of the lonely, the wild and picturesque, and the botanist, than it will the geologist. Still there is the slate to study; the whole walk is upon slate; and the many changes in its mineral character, the effects of former sea action upon it, as shewn in the ancient cliff, which in some places advances boldly on the shore, and in others retires inland, and as compared with the like action on the sandstone, will form interesting subjects of inquiry. Abrupt changes in dip will be noticed in the slate, at the north side of Glen Catacol. With the exception of a few singular inversions, due, probably, to dikes, it maintains a pretty uniform dip and inclination, 40° E. of S. at about 40°, till we approach Glen Catacol, north of which an anticlinal seems to occur, owing to an advance of the granite in the direction of the bay. On one side the slate rises up towards the granite, as if thrown off from it, the dip being 65° W. of N. at 53°; on the other it dips E. and S.E., at about 30° to 40°. But this dip is not maintained; it is sometimes directed towards the granite, in other places "end on" to it, or southwards, and in others again off the nucleus, or nearly west; but there is never exhibited that mantling stratification around the nucleus which is usually seen in granite tracts. The entrance of Glen Catacol exhibits some fine terraces. On the south side there are two; one about 40 feet in height, with which a mound rather than a terrace on the other side corresponds; and a lower terrace about 20 to 30 feet high, which reaches down to near the sea. With this a terrace on the north side corresponds, beginning near the farm-house, and sweeping round eastward. The front of the higher terrace is cut into promontories and bays, indicating the action of water upon it after its deposition; all the terraces consist of alluvial matter, not of rock. When the sea stood at its former level so often alluded to already, the plain in the centre of the glen must have been entirely under water—there is but a thin soil over rolled gravel through all its central part. On the north-east part of the glen, and out eastwards over the mountains towards Tornidneon, some very interesting junctions are to be seen; but we must pursue our way now towards Dugarry.

It is only on this extreme north-western border of the slate tract that the micaceous character which prevails in Cantire is assumed by the slate. Near the Free Church, which stands on a lovely picturesque spot in North Thundergay, a fine mica slate, banded with quartz, occurs; east of this there is much chlorite slate, and one mile west of it, near the secluded burying-ground, blue argillaceous slate, mixed with the chloritic variety. But generally there is on this side a greater tendency to the micaceous character. On the Imochair shore, and towards Dugarry, the ordinary character has become again established. The slate ceases, and the old red sandstone begins to form the front of the cliff a little to the west of the mouth of the Iorsa; but the junction is not seen. The stratification appears to be conformable.—There are comparatively few dikes on this coast. A little east of the Free Church there are two, each of which appears to bifurcate. On the Penrioch shore a great bed of greenstone, 100 feet broad, occurs. We did not notice a single bed either of porphyry or claystone. Many boulders of granite occur; northwards they are of the coarse-grained variety, but of both varieties on the South Thundergay and Penrioch shores. A few were estimated at as much as twenty to thirty tons weight.

As the coast from South Thundergay southward is somewhat monotonous, we would recommend the tourist, if the day be clear and steady, and he has had an early start, to diverge inland at South Thundergay, visit the lovely and secluded corrie and tarn, called Corrie-an-Lachan (hollow of the lake), ascend Ben-Varen from the north, pass along its eastern ridge, and descend upon Iorsa waterfoot, through Glen Scaftigill. He will be delighted with the wild scenery, and will notice many objects of geological interest.

The terraces already noticed (Art. 18) at the mouth of the Iorsa, in front of the Duke of Hamilton's shooting lodge, are very striking—by far the most remarkable in Arran. They are the remains of a vast accumulation of detrital matter which once filled the whole of the deeply-embayed area at the mouth of the river, and was most probably deposited as a terminal moraine to two glaciers, which moulded the sides of Sal-Halmidel, and united their streams near the head of the bay. The seaward front and inland extension would be originally nearly half a mile each. The detritus backs against the sides of the sloping hill all around, and is seen to rest against the natural rock, fronting the sea, outside the bay, at the same level with it, and worn like it into the existing cliff. This effect is especially remarkable on the west side. Here the slate, the old red sandstone, and the detritus form, in succession towards the east, the front of a cliff, at first of nearly uniform height, and then gradually declining eastwards, all the three portions presenting exactly the same evidence of sea action. They form, in fact, an unbroken line of old sea cliff, carved out on all three alike when the land was rising, in the manner already pointed out (Arts. 18–20). Mr Robert Chambers has stated it as his opinion (Edin. New Phil. Jour., N. S., vol. i., p. 103; and Edin. Papers, "Ice and Water," p. 20) that in this case "the moraine has been thrown down upon the terrace so as to fill up its deep angle;" and he hence, of course, draws the conclusion that the moraine was deposited after the terrace was formed. The appearances, as we have above described them, seem to us strongly to favour the opposite conclusion, that the moraine was thrown down before the last elevation of the land, and that the sea cliff was carved on it, as on the rock. The contact of the slate and sandstone is not seen in the cliffs, nor on the terrace in front, where worn masses of slate in front of the slate cliff, and of sandstone, in front of the other two portions, all *in situ,* rise up through the shingle. The foundation rock all round eastwards by the lodge and east side of the bay is sandstone. The detritus was cut into the existing forms, by the joint action of the sea and river; the mounds present in front rounded promontories and semi-circular bays, and on the ascent terraces at several successive heights. An isolated mound in front of the lodge was removed in order to open up the view eastwards. Among the detritus there are beds of fine sand, much sought after by gardeners, and often carried to the mainland.

There is no hotel in this part of the island, but very comfortable quarters can be had at a farm-house in the hamlet of Auchincar, a short distance east of the lodge. The tourist can then easily find his way to Brodick, or he may pass along the shore by Mauchrie waterfoot to Tormore and King's Cove. From Dugarry to the Tormore pitchstones, or to the Tormore Stone Circles, is a good hour's walk. Ben-Varen appears to best advantage on this walk; and we often pause upon our way to turn round and gaze on its stupendous mass, and to wonder how it came there. A quarter of a mile from the shore there is a wooden footpath across the Mauchrie. About half a mile from the sea, between the farm-houses of Mauchrie and Tormore, carboniferous limestone, with *producti* and encrinal stems, occurs as a bed in the sandstone of Mauchrie water, and is very similar to the Corrie beds, and, like them, the limestone is not far from the base of the series. In Mauchrie burn, close to Mauchrie farm-house, the Old Red beds are seen; but no precise line can be laid down for the boundary. Glen Iorsa opens on this part of the coast; and here the number of granite boulders greatly increases—the proportion being now in favour of the fine-grained variety, which exists throughout the whole of Glen Iorsa. From Iorsa waterfoot the coast section presents first old red sandstone and conglomerate, and then carboniferous sandstones. We observed scratched boulders about Auchincar, and masses of old red sandstone, which in this situation are not travelled, also striated.

EXCURSION VIII.

By the Interior of the Nucleus.

65. A walk across the interior of the granite nucleus, from Loch Ranza to Iorsa water-foot, will be good work for a long summer's day; or the geologist may return from Auchincar to Loch Ranza by this route. We preferred the former course, passing up Glen Eais-na-vearraid by the base of Tornidneon. The granite junction here is very celebrated, having been often referred to in the discussions between the abettors of the igneous and aqueous theories—the Huttonians and Wernerians. In our view the appearances are strongly confirmatory of the intrusive character of the granite. About a mile south of Loch Ranza we leave the road and pass up to the base of Tornidneon, keeping on the west side of the stream. Tornidneon is a bold and finely-shaped mountain, about 1500 feet high, which rises on the east side of the gorge through which the fierce blasts already alluded to often come down, while Meal-Mhor rises over it on the west.* Through the gorge a torrent descends into the head of the loch, having cut a deep channel through the hard layers of slate. The place is wild and picturesque; at some points great caution must be used in selecting our footsteps. It was through this gorge that a glacier, which filled the wide mountain glen beyond, debouched upon the shore of a former period, and threw down into the bay the mass of detritus, which shoaled the water in this sheltered situation, and made the after work of the river comparatively easy. The river and tides have re-arranged the materials, and rolled them into a

* Tornidneon—the mountain of birds' nests; Mëal or Müil-Vawr—the great, round-topped mountain; Moil, Muil—whence Mull—is a round head, and is applied to mountains and headlands. The Gaelic language does not recognise diphthongs.

perfect marine shingle; but doubtless this was the origin of much of the debris which has filled up so large a part of the bay. The shingle bank on which the castle stands has very much the appearance of a terminal moraine; and it is just in such a position that a glacier like that supposed would for a long time throw a moraine down. There are traces of a former terrace along the sides of the bay, backing against the mountains, in the position which lateral moraines would occupy. When we said that the shingle bank has kept its place for 2,000 years, the reference was to the establishment of the existing levels, not later than the time of the Roman invasion.

66. At the upper end of the gorge, where the open glen begins, we are upon the junction in the river bed. Many fine branching veins of granite run into the slate, narrowing to threads as the distance increases, the granular structure becoming, at the same time, more minute; in some places granite bands are interstratified with the slate. The slate is also penetrated by quartz veins, of which some are parallel to the layers of slate and others intersect them at various angles; in the former case, they conform to all the contortions of the slate. In some of the larger veins the granite is coarse-grained; but usually a more compact or confusedly crystalline strip separates these from the slate. The appearances are finely exhibited in the front of the mountain, which is cut steeply down, and shows the veins in section. It is this circumstance which renders the junction here by far the finest in the island. "The whole mountain," as Murchison and Sedgwick express it, "abuts against the granite, which moulds itself into the broken edges of the slate, and runs into it through the gaps and fissures." Large veins are seen to emanate from the granite below, and pierce the slate in a slanting direction towards the summit. There are also many fine veins and numerous alternations where the two rocks approach; and here the usual changes are well exhibited—nowhere is metamorphic action more decided. The whole neighbourhood, indeed, is well worthy of careful study. It

will be found very instructive to ascend the hill along the plane of contact, and examine the various junctions. Many appearances equally striking are also to be seen across the hill-tops between this point and Glen Catacol, affording ample scope for a separate walk. We may ascend from the Tornidneon junction, and return from Catacol by the shore; or first visit Catacol from Loch Ranza, examine the terraced ridges at its mouth, and then ascend by the granite junction at the north-east angle of the glen. The descent upon Loch Ranza from the high platform is not difficult. The walk affords many grand views, particularly late in the afternoon.

67. From the Tornidneon junction southwards the walk is wholly on granite, and its chief interest, apart, of course, from its picturesque attractions, consists in examining the two varieties, the coarse and fine-grained, their relations and respective limits, and the dikes which traverse them. Few walks, indeed, in Arran, are more delightful on a fine day; the peaks and ridges are seen in new and grand aspects, enhanced by the solemn stillness and desolate character of the scene. The silence is broken only by the wild scream of the curlew roused from her nest, or the answering call of her mate aloft, by the whirr of the startled wild-fowl, or sudden bound of the red deer from his lair. From Loch Ranza to Dugarry there is not a single human habitation, not even a herd's shieling; and during the ramble of a long summer's day we shall meet only the wild denizens of the rock or moor. The red deer seek chiefly the shelter of the lower valley of the Iorsa; and it is thence by the wild heights, and rocky knolls towards the watershed that the sport of "stalking" is occasionally followed. The deer are not very numerous; but they have been put into some other covers, and the small American species has lately been placed in the woods about Brodick Castle.

The botany of this district is meagre, as is usually the case with open uniform granite tracts; the only good plant is the rarer species of *Pyrus,* which is found in rocky places on the stream, and at the base of The Castles. The junction of the fine granite with the coarse generally takes place where the gradual slope ends, and the steeper climb to the mountain tops begins. An important service would be rendered by tracing this line along both sides and observing the appearances—whether they are similar to those already noticed by us on the south side of the district (Art. 14); whether veins pass from the fine into the coarse; and whether dikes pass from one into the other continuously, or are broken abruptly off (Art. 6). It would be almost necessary to "camp out" two nights running, and it is not easy to accomplish this in a climate so uncertain. Facts have been already stated (Arts. 6, 14, 29) which render it highly probable that the dikes do actually pass continuously through both kinds; but nowhere has such actual passage been observed by us at the line of junction of the two granites.

Having reached the summit-level on our ascent from the north, we are at the curious little tarn Loch-an-Deavie, which, when it rises about eighteen inches above its level in dry weather, discharges both ways, as already noticed; another nameless tarn has the same curious relation to Glen Catacol (Art. 2). But neither Loch Dhu, or Loch Tanna, has a similar position. The water from them passes southwards.

Cross-Moor is the high granite mountain between the water-shed and Ben-Varen. On its south-east side, a whin dike was noticed four and a-half feet broad, ranging about N. 26° W.; it is in the fine granite, and was traced downwards a good way towards the Iorsa, which it would seem to cross perpendicularly. It consists of greenstone, partly fine-grained, and partly porphyritic and amygdaloidal, with much iron. Along the side of the hill towards the S.W. another dike of a similar trap was seen; the width 10 feet, and range N. 35° E., or at an angle of 61° with the other dike. From this point to the head of Loch Tanna indications of two other dikes were observed, from which it was concluded that they had a range of W.N.W., and N.W. by N.—We may now examine the great beds of china clay under Ben-Varen to the N.W. of Loch Tanna;

then ascend Ben-Varen, pass along its horse-shoe crest, and then, having crossed the junction of the granite with the slate, which here ranges W. 25° S., or nearly "end on" to it, dipping S. 25° E. at a high angle, descend upon Imochair; or we may pass on southwards by the base of Sal-Halmidel to the mouth of the Iorsa, the two routes here uniting at Dugarry. We are then near comfortable quarters at Auchincar; or, by previous concert, a vehicle may be in waiting to carry us to Brodick—ten miles—very welcome it will be after so long a march.

EXCURSION IX.

Loch Ranza to Corrie.

68. The inland route from Loch Ranza to Corrie, a distance of nearly eight miles, presents a few objects of interest. A walk of three miles by the high road, which winds screw-like along the hill-sides, by the edges of the deep glens, brings us to the summit-level of Glen Chalmidel, and the watershed into North Glen Sannox. The views of the northern front of the ridge of The Castles and the Sui are very grand from this point. A few hundred yards below the summit, as we pass down into Glen Dhu, granite appears by the road-side; and in the river bed, a little farther, a junction of the slate and granite is very well exposed. A glance at the map will explain the occurrence of the granite here. The principal junction is at the point where the river comes close in under the bank, along which the road passes. Here there are some peculiarities well worthy of notice. The two rocks are seen to come close up against one another, most distinctly contrasted; but there are no veins emanating from the granite. This rock is coarse-grained, much coarser than in any other junction we have noticed, yet it is more compact than the coarse-grained granite usually is, and, though hard, brittle. About one hundred yards down the stream a vein of fine, compact granite, five feet broad, traverses the slate; and thirty yards farther, another vein, one foot broad, bluish coloured, fine-grained, and very brittle. The entire mass of slate between this lowest vein and the main junction above is much altered; it is extremely hard and tough, contorted in the laminæ, and being much veined with quartz, is assimilated in aspect to a granite rock. Below the lowest vein it speedily assumes its usual appearance. Mr

Ramsay was the first to notice this interesting junction; and it is well described by him. Several other junctions may be seen in the branches of the main stream.

A little farther down the glen, and less than one mile from the bridge, some well-marked glacial striæ are seen on the road-side, upon a mass of slate *in situ*. They cross the laminæ, and are directed nearly due east and west. The slate also is rounded off.—On the hill to the north, some great blocks of coarse granite rest upon the slate. These travelled blocks reach a good way up the south front of the coast range; but in crossing, on one occasion, along the summit, keeping generally near the edge of the cliffs, we did not find a single block. It would seem, therefore, that the few and comparatively small blocks on the north shore, must have worked round, from either side, under the force of the waves. Slate was found everywhere on the summit, and far down upon the steep fronts, where the junctions take place.—Some curious varieties of slate will be observed in the river bed, on both sides of the bridge. A short distance below the bridge, we leave the slate, and, after an interval, come on coarse conglomerate, no junction being visible.

Many years ago, a large population, the largest then collected in any one spot in Arran, inhabited this glen, and gained a scanty subsistence by fishing, and by cultivating fertile plots on the sunny hill-sides. In 1832, the whole of the families, amounting to 500 persons, were obliged to leave the island, but were furnished with the means of reaching New Brunswick. They formed a settlement at Chaleur Bay, which became very prosperous. Garden enclosures, tree-clumps, solitary fairy thorns, and ruined wall-steads, still remaining, give a melancholy interest to this secluded glen.

Near the summit level of the road between North and South Sannox, and a little to the north of Sannox House, the sandstone strata, on the east side of the road, are marked with striæ, rendered somewhat fainter than those on the slate by the effect of decomposition. Like these they are transverse to the laminæ, and are directed W. 4° S., that is, in the direction of the ridge of Suithi-Fergus. If we assign their origin to ice, then we must admit that the moving masses advanced in parallel courses, in both cases, from high ground on the west.

We are now within a few yards of the spot where the finest view of Glen Sannox is obtained, and to which a former excursion brought us. Taking another survey of the grand scene in its evening aspects, we press forward to Corrie, to catch the Clyde steamer on her way to Brodick.

EXCURSION X.

To Holy Isle.

69. Make choice of a bright and quiet day, and leave Invercloy for Lamlash by the mail steamer from Ardrossan. On reaching Lamlash, a boat must be hired to carry you to Holy Isle. The pier having fallen to pieces, the steamer does not stop, and there is no one permanently resident who could be signalled to come off. A few hours will suffice to examine the island, and the charge for a boat is moderate—from 6d to 1s per hour. The basis of the isle all round is sandstone, which rises to the height of 100 to 150 feet; the rest of the mountain, to the height of 1,020 feet, is composed of claystone, so that this rock has the thickness of about 850 or 900 feet. On part of the east side, however, it is much less than this, as the sandstone rises there much above its usual height. The rock is of igneous origin, a member of the trap family, and varies in structure from a soft claystone to a compact felspar; the harder varieties are called clinkstone. As in most traps, the prevailing form is prismatic; and on the east and south-east sides of the island, columns of great length appear in the high precipices. But the schistose structure is also common; and, as in cases already mentioned at Corriegills, both structures occur in the same mass, the slaty fracture being at right angles to the axis of the prisms; the ends of the prisms first divide into laminæ, and the mass gradually assumes the slaty structure. The weathering of the rock is remarkable, and extends to a considerable depth, presenting successive concentric zones of different colours, which have a very pretty appearance in many specimens. The slaty structure itself, in the case of this rock, seems to be but a step in the process of decomposition.

Various masses of greenstone occur, both as dikes and beds; one on the south-east side of the island presents remarkable alternations with the sandstone, and alters its structure. The felspar rock itself is seen in one place on the east side to intersect the whole body of strata, and to be connected above with the overlying mass of the same rock. Here, also, there is a very interesting vertical vein or dike, intersecting an oblique trap vein, and showing in the centre fine black basalt, containing zeolites and glassy felspar; on each side of this there is a lamina of a black substance, intermediate between basalt and pitchstone, and at the outer surfaces a coating of vitreous pitchstone one-tenth of an inch in thickness. These substances, indeed, are all of one origin, the variations found here, as well as in other places, being due merely to different rates of cooling in the once fused mass (see Art. 52). The same oblique vein is intersected by a vertical vein of concretionary trap. Dr MacCulloch gives a drawing representing the appearances (*West Isl.*, vol. iii., pl. xxiv., fig. 2).

The granite blocks on this island have been already noticed (Art. 19). On the west side, near the landing, there is a raised beach or shelly deposit thirty feet above the present sea-level. Here also once stood a fortress, built in the middle of the 17th century by Somerled of the Isles, of which no trace now remains. Adders abound upon the island, but their number is said to be diminishing.

70. The visitor must see the hermitage of St Molios. It is situated a mile south of the landing-place, and is a natural excavation in the old sea cliff. The conjectural account of St Molios usually given is quite inaccurate. The following is the correct legend, kindly made known to me by the late John M'Kinlay, Esq., of Bonnington, near Edinburgh, formerly of Rothesay*:—"The legend of St Lasrian, or Molassus, is well known to Irish antiquaries; it was printed in the Bol-

* Mr M'Kinlay's extensive and accurate information in regard to the history and antiquities of his native county are now in a great measure embodied in Mr Reid's detailed and excellent "History of the County of Bute."

landists' collection, *Acta Sanctorum.* He was born in Ireland A. D. 566, educated in Bute by his uncle, St Blaan, returned for some years to Ireland, and afterwards, when yet only twenty years of age, retired to an island in Scotland, where for some years he led the life of a hermit—probably in the cave which is still pointed out as his in Lamlash Isle, *i.e.*, 'Isle-a-Molass,'* later named the 'Holy Isle.' About the year 614 he was elected Abbot of Leighlin, in Ireland, and was afterwards made a bishop, and Apostolic legate to the Church in Ireland. He died in the year 640. The inscription on the roof, in Runic letters, has no reference to Molassus. The words, '*Nicolas hann raisti*'—'Nicolas this engraved'—are Norse or Icelandic, and clearly refer to a Norwegian hermit who resided here at the time when the Northmen ruled the Western Isles, or about A.D. 1100. Wilson (*Arch. of Scotl.*, p. 531) identifies this hermit with a bishop of Man. He would make the inscription refer to the excavation of the rock. But this has clearly been the work of the sea; and, besides, the cave was the abode of St Molassus at an earlier period. Mr Wilson gives some other fragmentary Runes. A small cell or monastery was erected in connection with the hermitage, apparently by Reginald de Insulis, between 1206 and 1212, probably in connection with the monastery of Saddell, in Cantire, founded by him, and to which he granted lands in Arran. It seems to have been an abbot of this small monastery, whose tombstone, bearing his chalice and pastoral staff, but without any inscription, is still extant in the ruins of the ancient burying-place and chapel at Clachan glen, and which is popularly called St Molios' grave." The traces of this small cell or monastery were till lately to be seen north of the cave, marked by an old hawthorn, and beside a burying-ground, which was long used as the chief place of sepulture for Arran, till the loss of life by the upsetting of a boat led to a discontinuance of the

* Lasrian and Molassus are radically the same. "Las," meaning "light," was probably the proper name; the prefix "mo," and affix "rian," mean "very" and "good," expressing approval or endearment.

practice. A pure spring of water close by the cave—the saints' well—was long famous for its supposed healing qualities.

Returning to Lamlash, we may examine the ruins of the old parish church of Kilbride, on the east side of the Brodick road, N.E. of Lamlash, which, though without any architectural beauty, has some monuments of interest. Of these Mr M'Arthur gives an account in his History of Arran; he also figures some of the monumental stones. The summit-level of the road to Brodick is marked by four granite blocks enclosing a circular space, used as a place of ancient sepulture, to which allusion will be made in the sequel, in connection with an account of the Tormore Circles. A few yards south of it there is an upright stone of coarse conglomerate, about four feet high, and others near it now prostrate, all apparently portions of a former circle.

EXCURSION XI.

To Windmill Hill and Ploverfield.

71. The Windmill Hill is conspicuous from all parts of the Brodick coast, as the high narrow crest of a long ridge which divides Glen Shirag from Glen Cloy. It may be readily ascended from any side. If we pass along its north front from the String road, we shall see the pits from which shell limestone, like that at Brodick Church, was formerly quarried; but the ascent along the south-eastern front will best expose to our view the structure of the hill. We pass up the lovely banks of the Cloy burn, fragrant all the way with blossoms of the choicest wild-flowers, till we reach the bridge leading to the old mansion of Kilmichael. This is the seat of the Fullarton family, proprietors of an estate in Arran—the only portion of the island not possessed by the princely house of Hamilton. Of part of the property, Whitefarland, on the north-west side of the island, the Duke of Hamilton is lessee. It is not exactly known in what way the family first came to possess lands in Arran; but there seems little doubt that they acted as stewards in the island in the time of Bruce, and that the lands of Kilmichael were granted them in his time; for, in the charter granted by King Robert III., in 1391, and which is still extant, bestowing the lands of Ergwhonnyne or Strathwhillan, Fergus Fullarton is styled "of Kilmichael." The property of Whitefarland was acquired by marriage. The name appears to have been at one time MacLewis or MacLuoy, now only preserved in the name of the glen in which Kilmichael House is situated. The family is one of the oldest in the county, and seems to be of the same origin as the Fullartons of that Ilk in Ayrshire; and probably also as the Fullartons

of Mindmock, heritable coroners of Bute, called also MacCamie, that is, son of James or Jamieson. It is not known in what way the name MacLewis or MacCluoy came to be applied; but such designations are common in Highland families, as distinguishing the branches of a clan. The Fullartons were heritable coroners of Arran, as the MacCamies were in Bute.*

The front of Windmill Hill is formed, through about half its height, of columnar felspar porphyry, similar to that of Dun-Dhu, already described (Art. 41). The columns are four, five, and six-sided, with flat jointings; the pillars lean in various directions. The junction with the sandstone below is nowhere visible. About the middle of the front of the hill, a mass of altered sandstone, 12 feet wide, is imbedded in the porphyry. Close to this is a whin dike, running N. 10° E. Several others traverse the ridge of porphyry—another example of the posteriority of the common trap or whinstone to all the rocks of Arran. The plane of contact between the sandstone and porphyry gradually ascends westward, and, on reaching the S.W. shoulder of the hill, we find a wedge-shaped mass of sandstone connected with the body of this rock below, and apparently separating the porphyry from the granite which immediately succeeds. No true contact of these rocks is observable. Sandstone, slaty pieces of porphyry, granite sand, and bits of granite are seen lying about, mixed up confusedly, and no line of demarcation can be laid down. Presently the hill-side shows granite only. Eastwards all the high ridge of Windmill Hill is porphyry; sandstone is seen at the north base, but no contact is visible. The base of the porphyry is an intimate mixture of felspar and quartz, of a gray or bluish-gray colour, with imbedded felspar crystals, bits of quartz, and occasionally well-formed crystals of this substance. We have already (Art. 11) described at sufficient length the relations of the sandstone to the Ploverfield granite, to this porphyry, and the syenite of the hill-sides westwards, into which the granite seems to shade off; and need not

* The history of the family, so far as known, is given by Mr Reid in his "History of Bute."

now recapitulate. The series forms an interesting study. The granite, porphyry, and syenite, are all posterior to the sandstone; but we have as yet no means of knowing the age of the porphyry in relation to the granite. The common greenstone of the dikes, and detached knolls upon the plateau southwards, is the newest of all these igneous products. The occurrence in this sharp-crested hill of an erupted rock like porphyry, must be explained as we explain the existence of granite in mountain peaks, or basaltic lavas in the mural precipices of Mull and the Giant's Causeway.

72. Passing round the heads of Glen Dhu and Glen Cloy (Art. 11), we come over the edge of high sandstone cliffs bounding Glen Cloy on the south. On a grassy ledge under the western part of these cliffs there occur two remarkable dikes, producing a highly interesting change upon the sandstone. A large body of this rock, between the two dikes, is altered to the state almost of quartz rock, and beautiful crystals of amethyst are developed in it; quartz crystals, both colourless and with a slight tinge of yellow, also occur. The dike on the S.E. side is a brick-red porphyry, resembling that on the Corriegills shore; it appears also in the cliffs above, where, by wearing, a fissure is formed upon it; the dike on the N.W. side is of greenstone. There is a third dike or mass of trap outside the porphyry; but neither its relation to the others, nor of these to another, can be well made out. The locality altogether is fully as interesting, on account of the striking metamorphism, as any in the island.—The isolated high round knolls, called the Sheeans, over the head of Glen Cloy, are trap. Passing from them to the head of Lamlash, or Bein-Leister Glen, we meet with a few low knolls of like composition, but no body of overlying trap. Lamlash Glen affords a fine section of the carboniferous series: red limestone and red marl at the top of the glen, and lower down two other considerable beds of limestone, the last being about a mile and a-half from Lamlash. They contain the usual fossils, and occur amid massive sandstones. A short distance below the lowest limestone, a dike of

felspar rock, or quartziferous porphyry, fifteen to twenty feet wide, crosses the bed of the river, ranging nearly N. and S. The sandstone is greatly altered by it. Between this point and the alluvial plain there are several veins or beds of claystone and greenstone, breaking through the sandstone of the river bed.

We can easily pass from this point into the lower part of Moneadh-Mhor glen. At the opening of this glen, in the bed and banks of the stream, and of the lead of water connected with the large mill adjoining, there are two great beds of pitchstone. They are finely exposed, and exhibit very strikingly those transitional appearances already alluded to as marking the relations of this rock to hornstone and claystone.

Associated with the pitchstones there are claystones, hornstones, quartz rock, and porcellanite. Hornstone, that is, chalcedonic or jaspery chert, seems to be the link between pitchstone and claystone. Hornstone and pitchstone are both almost entirely siliceous; the difference consisting in the colour and degree of toughness arising from a slight change in composition, or variation in the rate at which they cooled, or from both. By this change pitchstone passes into hornstone. In this hornstone a great many light-coloured spots with dark centres are gradually developed; and bands of this variety succeed the common hornstone. Next to this there follows a hard quartzose claystone, and the series terminates in the common claystone of open texture, like that of the Corriegills shore already described. The spots or specks are minute spheres, most probably of felspar or quartz, or of both, and have probably originated in the manner suggested at the end of Art. 39. They present a close analogy with the spherulitic claystone of Corriegills; but the radiated structure of the latter does not exist here. Even the larger spherules of the pisolitic hornstones do not exhibit this structure. The quartz-rock at the upper end of the higher pitchstone vein is probably only an altered sandstone. The porcellanite alluded to above is a white substance, varying from a dull earthy aspect to that of a white enamel. It occurs of considerable thickness along the outer surface of the pitchstone, and is clearly due to a decomposition of this rock. The incipient stages of this decomposition shew a structure in the rock which otherwise we should not have suspected. The surface is traversed by a series of wavy lines, conforming to one another throughout. This indicates a laminar arrangement within; and it is along these lines that the rock splits completely up in the advanced stages of decomposition.

The relations of these various igneous products is further illustrated by another dike, which occurs a little way up this river. A great vein of claystone crossing the bed of the stream shows a distinct passage into jaspery hornstone. The jasper and chert veins in the Tormore pitchstone, with associated claystones and basalts, place these same relations in a very clear light; and when we view the appearances which they exhibit in connection with those now described, we cannot hesitate to admit that all these products of fire blend into one another, the varying aspects which they assume being due to a slight change in chemical composition, in molecular arrangement of parts, or in the rate at which they were consolidated from a state of igneous fusion.

EXCURSION XII.

To Ceim-na-Cailliach and The Castles.

73. Excursions in Arran may be varied indefinitely according to the taste and objects of the tourist or student. We have indicated a few best fitted to shew the most striking geological features of the island; and as we have now described the different formations and the most remarkable appearances which they exhibit, we shall only mention shortly the chief remaining objects of interest in the northern division, which may be visited in two excursions. We must then carry the student on three excursions—two by the South End, and one to King's Cove and the Tormore Stone Circles.

The principal object in our present excursion will be to look for "the black crystals," as the smoke-quartz crystals found in the granite are called. Their chief repository is the north side of the Suithi-Fergus ridge and that of The Castles. They are found in the coarse-grained and rapidly disintegrating granite of the great northern ridge. A guide should, if possible, be secured; let him carry a pick-axe to open up the disintegrating rock; it will also be well to carry a pretty heavy hammer; considerable blocks will then be easily broken. The crystals may also be found by searching the debris, where each little stream bursting out of the rock, forms a talus on the dispersing of the water. Occasionally beautiful specimens of granite may be found with the ingredients crystallized.

The following analysis of the felspar of the coarse granite has been kindly made for me by Mr Magnus M. Tait, F.C.S., assistant to Dr Anderson, Professor of Chemistry, Glasgow College:—

Silica, ...	63·70
Alumina,.	20·02
Potash, ...	12·33
Soda, ...	1·71
Oxide of Iron, ...	1·28
Lime, ...	0·89
Magnesia, ...	trace.
	99·93

This analysis shews that the felspar is an orthoclase or potash-felspar; and that it consists of one equivalent of tersilicate of potash, and three equivalents of tersilicate of alumina; its formula is $KO3SiO^2 + 3\ (al^2\ O^3\ 3SiO^2)$.

We pass up by the N.E. angle of Glen Sannox, and then diagonally along the back of the ridge of Suithi-Fergus. Here the whole length of the lofty ridge on the south side of Glen Sannox is before us. It terminates to the left in three peaks, to all of which the name of Maidens'-pap or Cioch-na-Oich is assigned by the Royal Engineers; the heights they have found for these are as follows—peak to the N.E., 2,541 feet; to S.E., 2,172 feet; to S.W., 2,687 feet. The long jagged ridge ends towards the right in the two lofty summits of North and South Goatfell.

Ceim-na-Cailliach, or the Carlins' Step, is an immense chasm or gash in the ridge, overlooked by granite walls several hundred feet in height. The interior of the fissure can be easily reached by entering laterally at a pretty low level on the north side. We find it to be merely a whin dike worn down to this great depth below the containing granite. The rock is a dark-coloured, fine-grained greenstone of loose texture; it exhibits the concentric spheroidal structure so often alluded to as characteristic of common whinstone. There is no trace of any other rock; and we cannot understand how it has come to be so often said that this dike is pitchstone. The dike and chasm pass down into Glen Sannox, but it is very unsafe, if possible, to descend by this way. The view is very wild and grand. In order to ascend to Castle Abhael, we must pass north down the chasm, and then ascend towards the west. Here the glen and ridges are seen in a new aspect. When the morning mists are hovering around Goatfell, and rolling into the depths of Glen Sannox, now hiding and now revealing the mountain with the ridges and peaks adjoining, the scene from the summit of Castle Abhael is extremely grand.—Various ways are open to us by which to return. We were once surprised here by a thick fog with heavy rain, and guided partly by a compass, and partly by keeping as close as possible to the junction of the coarse and fine granite, we found our way safely along the edge of the great circular ridge which runs from the north-west base of Castle Abhael, in the direction of Glen Dhu, at the upper part of North Sannox. But the convenience of our companion, with his load of "black crystals," must now be consulted in regard to our homeward route.

EXCURSION XIII.

Glen Sannox and Glen Rosa.

74. This is a favourite excursion with visitors to Arran; many will devote a day to it in preference to others which are less easily taken from Brodick as a centre. But there are comparatively few objects of geological interest, and we shall not therefore enter into any detailed description.

There is an interesting old cemetery at the entrance of Glen Sannox. It is all that remains of a chapel dependent on Kilbride Church, and dedicated to St Michael. A rude image of the saint is to be seen upon a stone built into the outside of the wall. The house was probably connected with the Abbey of Kilwinning, to which Sir John Monteith, Lord of Arran, granted, in 1357, the lands of Sannox and patronage of the churches. The barytes mill, which till lately sadly marred the solitary grandeur of the scene which opens as we reach the plateau at the mouth of the glen, was entirely removed two years ago by order of the late Duke of Hamilton. It was erected here to grind the sulphate of barytes, or heavy spar, raised from veins which traverse the old red sandstone. These are seen in the bed of the stream, running in a direction nearly N.E. and S.W., and dipping variously; they appear also on the hill-side southward. The junction in this burn is not well seen, but interesting junctions occur under the base of Cioch-na-Oich. By the bank of this fine stream of crystal water, rushing over its bed of granite sheets and granite sand, amid huge rolled masses of the rock, is a delightful walk. Near the head of the glen we may diverge a little to the right, and examine, at this lower level, the dike and great chasm descending from Ceim-na-Cailliach. Following it southward on its line of bearing we trace it entering the granite precipices in front of the base of Cior-Mhor, and passing on over the col, in the direction of the axis of Glen Rosa. The safest pass into this glen is at the western side of the col or ridge joining Cior-Mhor to the base of Goatfell. In the hollow up which we pass we have a whin dike beneath our feet, and granite walls on either hand—a pathway, in fact, has been formed by nature in the disintegration of this dike. On reaching the summit we observe a great dike, most probably a continuation of this one, ranging right up the front of Cior-Mhor. When one reflects upon the mode of origin of these two rocks, granite and greenstone, it strikes one with wonder to perceive the curious relations which they maintain, and the important part in the physical condition of the region which the dikes play. If the molten matter of greenstone had had full vent here and overflowed, the interior pressure would have been relieved, and the mountains of the granite nucleus would have stood at lesser altitudes. M. Necker goes so far as to suppose it quite possible that the excavation of Glen Rosa may have been determined by the great dike above noticed. In the lower part of the glen the river runs upon it, between high granite walls, for a long distance; and it crosses S. into Glen Shirag. It is thus by far the most continuous dike yet traced in Arran.

Up till the year 1822 this path was known as practicable only by shepherds, some of whom occasionally used it. But in that year two enterprising young ladies, Miss Alison and Miss Crooks, both from near Kilmarnock, but residing at Brodick, having arrived on a summer afternoon at the top of the ridge by passing up Glen Rosa, determined to try the descent into Glen Sannox, and return by the coast road. With great difficulty and loss of time they made good the descent, but were so late on arriving at Brodick that all the young men of the village had started off in parties, in different directions, to search for them. Their tale excited no small wonder.

If the climber so desire, he can easily pass into Glen Iorsa from the head of Glen Sannox, across the col between The

Castles and Cior-Mhor; he will find some good plants by the way, and from the top of the col reach the summit either of Cior-Mhor or of The Castles without difficulty.

A very interesting dike exists on the ridge between Glen Sannox and Glen Rosa. It intersects the high southern part of the ridge between the two glens, near the point where North Goatfell starts up from the level of the col. The dike here consists of green, or nearly black, pitchstone only, but the sides are not seen in contact with the adjoining granite, nor is the dike traceable downwards into Glen Sannox. It ranges, however, nearly due west across the head of Glen Rosa, a little below the col, and thence right up the face of Cior-Mhor, where it is found to consist of a central band of green pitchstone, with six other bands, three on either side. The first of these, next the pitchstone, is a quartzose claystone approaching to hornstone, with light-coloured spots in the base, apparently both of quartz and felspar; to this succeeds a band of hard claystone porphyry, outside of which on either side is a broad band of fine-grained greenstone with imbedded crystals, constituting the variety called trap-porphyry. The width of the vein in the first part of its course cannot be determined; in the front of Cior-Mhor it varies from 20 to 30 feet. The vein forms a ravine up which one can clamber to the summit. Above this ravine the claystone disappears, and the pitchstone central band widens from 2 or 3 to 5 feet, while each of the trap bands is about 4 feet. As illustrative of these curious relations, the remarks in Art. 72 may be referred to.

75. There is little else of much interest in Glen Rosa; the granite junction has been described already in our First Excursion (Art. 26), and the singular chasm just noticed will be seen as we pass along. The adder (*Pelias berus*) is often met with in the glen, on the dry parts of the path, and in places where there is a dry bottom under the heather; and the tourist in crossing these will do well to use caution. Again and again we have narrowly escaped treading upon them in such situations, here, on the North Sannox shore, in Lamlash

Glen, and other places. We have never seen them higher, however, than the top of Glen Shant Rock, 1,100 feet; nor are they as abundant on the west as on the east side of the island, on account, no doubt, of the shelter and warmth. They are found also in Holy Isle. The occurrence of these creatures thus on islands is very singular, yet part only of a great physical problem, like the existence at the present day of the tiger and elephant in Sumatra, or the former existence of the hyæna, hippopotamus, elephant, and other wild beasts in Great Britain; or the camel-leopard in Greece. Man does not often interfere for the dissemination of such creatures; yet a case of such interference has occurred in our own time. The late Mr Cleland, of Bangor, Downshire, a naturalist of some note, attempted to introduce into Ireland the creature of which we are now speaking, as well as the toad. Several individuals of both species were placed in his "policies," but in a few months one after another was found dead. It is very unlikely that such benevolent intentions were entertained by any naturalist of old towards the inhabitants of Arran or the Isle-a-Molass. The size of the adder in Arran seldom exceeds 2 feet; a common size is 18 inches, but we know a case in which a size of 2 feet 5 inches was reached. The adder is the only British representative of the poisonous group of serpents; but there is no authenticated case of fatal effects resulting from its bite in Arran or elsewhere in Britain. We have heard of more than one case of severe symptoms; but these were removed in a few days by an application of herbs boiled in butter. On one occasion only have I noticed aggressive attitudes in the adder. I was searching for ferns among the crevices of the limestone and sandstone beds on the north front of Maoldon, when a strange, prolonged h-i-s-s startled me; but I had no idea whence the sound came, or how it was produced; and it was a good many seconds before I caught sight of a large adder, within less than a foot of my hand, in an attitude as if to spring, with erect crest, open mouth, and tongue alarmingly visible—

"Attollentem iras et coerula colla tumentem."

On throwing in a small stone—for I have never felt inclined to kill an adder, having already a preserved specimen—the creature retired into its hole.

The Gaelic name of this glen is said to mean the Ferry-point Glen. This name "Ferry-point," as indicating that water may have filled the glen since man inhabited the island, is remarkable, especially when taken in connection with the anchor found in the glen. This was discovered, not on the hill-side, where peat is now cut, as stated in Art. 25, but on the level of the river above the parks, where peat was formerly cut. Art. 25 was printed off before the error of our informant was known to us; but it is clear that on evidence so slight as this, we cannot assign to a very great antiquity the use of iron anchors, nor a modern origin to the old sea-cliff which surrounds Arran; for it could only have been when the sea stood at that old level that Glen Rosa was occupied by the sea. That remarkable changes have taken place in Brodick bay in late times we know. The Duchess Anne built of Arran oak, and, in order to present it to the government, a vessel of considerable size, to the west of the group of ash trees near the present manse, and launched it there. And later still, about 120 years ago, it was usual to ship cattle, at high water, in the Cloy, from a natural pier of stone a little way west of the present smithy garden, north of the road.* But such changes as these of the recent period are of quite a different character from the occupancy of Glen Rosa by water: floods, tides, and storms may have produced the former—the whole island must have been raised 25 feet at least to lay dry the "Ferry-point" Glen. This name, it is true, may have reference to a lake; and certainly the present mound is very like the remains of an old lake barrier.

The burns of Arran are all beautiful and picturesque, but in different styles, according to the nature of the rocks across

* I am indebted to the kindness of Miss Brown of Invercloy for this and many other interesting facts regarding local traditions and antiquities.

which their courses lie. Those of the granite and slate tract are perhaps not the most beautiful; but this Rosa burn, especially in its lower course, and after it enters Glen Shant, surpasses in romantic beauty most of the bright streams of this lovely island.

76. Over the wood which covers the eastern end of the ridge on the south side of Glen Rosa, a curious meteorological phenomenon is often witnessed. When the wind is at S.W. or W., with a damp and warm atmosphere, a column of vapour is seen ascending from near the centre of the wood, remaining in a nearly steadfast position for some time, and then suddenly vanishing; to be again formed, and again as suddenly disappear; and this is repeated through a period sometimes of several hours. For some time we supposed that it was the smoke from a gipsy fire; but the explanation, no doubt, is, that currents of wind descending Glen Shirag on one side, and Glen Rosa on the other, produce, by the rapid out-draught into Glen Shant, a partial vacuum and vortex of light ascending air over the middle of the wood, which is completely sheltered; and by this relief of pressure a condensation of vapour takes place, just as a cloud is formed in the exhausted receiver of an air-pump.

We may mention now, when on the subject of meteorology, an interesting case illustrating the relations in which the different kinds of clouds stand to one another:—On a bright hot day, after rain on the day previous, a pretty rapid current of wind from W.S.W., with scattered masses of white fleecy clouds, prevailed aloft from early morning as a land wind; while a light east wind blew below as a sea breeze. The currents met, a little way out in the channel, off the Corriegills shore; and along a strip of sky, stretching from north to south, the fleecy cumuli of the S.W. current were arrested by the current from the east, and resolved into a broad band of cirri. By four o'clock in the afternoon the upper current had descended to the surface; and a strong breeze from the west, preceded by a whirlwind, immediately sprang up. Towards evening this gradually subsided into a gentle air. The day was one of those delightful Sabbaths to be enjoyed only in the rural districts of Scotland or the North of Ireland. The Free Church congregation of Brodick was observing the communion services; and a vast concourse of pious and most attentive worshippers was assembled from all parts of the island; one minister officiated in English in the church, while another conducted the service in Gaelic from a tent in the open field. Apart altogether from the solemn feelings awakened by the simple and touching service, the mere picturesque adjuncts of these scenes at Brodick will never fade from the memory of any one who has witnessed them.

EXCURSION XIV.

BY THE EASTERN SHORE TO DIPPEN AND LAG.

77. WE enter now upon a region whose geological structure is more simple, and strata less disturbed. The rocks we shall meet with are sandstone, felstone, and trap. The sandstone is rarely conglomerate, and has a greater admixture of argillaceous matter. Having been deposited as a fine sediment in quiet waters, and afterwards consolidated by the pressure of superincumbent ocean, aided by heat below, it was affected by a general movement and extensively fractured. Through the fissures thus formed streams of melted matter were forced up from lakes of molten rock in "the nether depths," and overflowed from hundreds of vents the surface of the sandstone. Cooling under varying local conditions, and having a slight local variation in its constituent parts, this matter became felstone in one part and trap in another, and differs from lava consolidated in the same way simply in this, that it cooled, not in the atmosphere, as lava does, but under a great pressure of incumbent ocean. Now, it must be remembered that the granite and slates of the north plunge beneath these sandstones, and must also have been pierced by the streams of liquid rock. This escape of molten trap, on a large scale, from beneath the granite, relieved that rock and the whole southern district from an enormous upward pressure; and hence the granite was not raised in a solid form; and hence, too, the contrasts between the northern and southern divisions of the island. Liquid matter poured out from beneath could not consolidate in the form of steep ridge or beetling precipice; the disruptions attendant on the elevation of the land, and the action of currents, would impress upon the surface most of the

existing forms; ice-action and ordinary disintegration would do the rest.

Sandstone is seen all round the shores of Lamlash bay. This bay, in fact—the best harbour on the Clyde—has been excavated wholly in sandstone; and such is the nature of the rock, that the bottom affords admirable holding ground, while the water is deep enough to float the largest ship. Holy Isle forms a breakwater, with safe and deep entrance at either side. "What stupendous might was exercised," says the Rev. Dr Landsborough, "when this gigantic mole was pushed up through rock and earth and water, and the elevated sandstone overflowed by a stream of melted porphyry! Behold the power and goodness of God! How many, after weathering the storm, and casting anchor under shelter of this mighty breakwater, have exclaimed: 'Thanks be to God, we are in Lamlash bay.'"* The shores exhibit in a striking manner the wearing action of the sea at a later period; dark trap veins intersect the shores in all directions, and some of these between Clachland point and Lamlash stand up like huge ramparts above the beach, shewing the extent to which the softer rock has been worn.

78. The distance from Lamlash to Lag by the mountain road, crossing the Ross and descending Glen Scorodale, is ten miles; by Dippen, twelve miles; or if we diverge to visit Kildonan Castle and the southern cliffs, fourteen miles. From Lamlash to Whiting bay the distance is about four miles. As the path by the shore has not much of new interest, we take the public road, which affords finer views of the scenery, and then pass down to King's Cross, at the S.E. angle of the bay. Here rocky ledges rise from a pretty pebbly beach, from which Bruce is said to have embarked when entering on his perilous enterprise of rousing the men of Carrick. A cross, no longer to be seen, commemorated the event; and close adjoining are the remains of a small fort, erected apparently for the purpose of defending this landing-place, one of the best in the bay and at

* Excursions in Arran, Chap. IX.

its entrance. There is a considerable body of coarse greenstone here intruded as a bed among the strata of sandstone, pieces of which are entangled in the greenstone, and altered to the condition of quartz rock. The greenstone is traversed by long continuous veins of a finer kind approaching to basalt, probably cotemporaneous. The rock is divided into large rhombic blocks, but the surface has no resemblance to a basaltic pavement. On the shore at Whiting bay there are multitudes of dikes, appearing in singular relations to one another—intersecting, bifurcating, uniting, and separating again, entangling and altering the sandstone, and so on—relations that it would be tedious to describe in detail. The shore is covered with granite boulders, one of which, near the south-east end of the bay, is larger than any boulder south of Brodick, with the exception only of the Corriegills boulder. It is remarkable that one so large should be found so far from the granite centre.

79. A little farther S. a fine glen opens to the west; this is Glen Eais-dale, at the head of which there is the highest waterfall in the island, bearing the name of Eais-a-Chranaig. The stream descends over trap and sandstone in two fine falls through more than 200 feet; and here, as throughout the rest of its course, the scenery is highly picturesque. On the south side there is a trap dike, 10 feet wide, cutting the sandstone and passing into the overlying mass of greenstone. The contact of these overlying masses with the sandstone shews a remarkable change. The rock has lost its red colour, and becomes a white felstone, slightly porphyritic. Farther down, the ordinary laminated character is gradually assumed. On a terraced bank to the right of the fall there are the remains of an ancient fortress, concealed amid wild wood, which appears to have been the strongest of all the forts of the ancient inhabitants, with the exception of that at Drumadoon. The wall, formed of huge slabs, without lime, is 25 feet thick and 90 yards in circuit. "From the situation of this strength, and the vast labour which has been employed in

rendering it impregnable to the attacks of an enemy, it is probable that it was used as an encampment by the early islanders for the security of their families in the case of invasion."* The local antiquarians point out the "hill of council" adjoining, where the chief men met, in case of alarm by beacon fires on the heights of Dippen or Dunfion; and tell you that there was a larger population in old times than now along these glens and sheltered slopes. On the brow of the terrace over the waterfall there are several slabs of greenstone marked with glacial striæ, directed nearly east and west, that is, in the direction down the glen. The hills westward are all of overlying trap, and in some places shew veins of pitchstone. This terrace, the lowest of several ascending steps, sweeps

round south-eastwards in a high cliff, and comes upon the shore at Leargie-beg, where for more than a mile southwards it forms sheer precipices of highly prismatic rock resting on sandstone, which it covers for a long way with its talus of fallen blocks—a very perfect counterpart, though on a small scale, of the great basaltic façades of Mull and the Antrim coast.

* "M'Arthur's Antiquities of Arran," p. 90.

Inland, the ground rises in successive terraces formed by the sheets of greenstone laid one over the other like the steps on a pyramid. Under the cliffs the scenery is wild and romantic, and the place altogether affords the best field in Arran for the study of the basaltic rocks. The peculiarities of soil and shelter have favoured the establishment of a few uncommon species of plants. On the level of the beach S. of the talus we first come on a great bed of claystone, very like the great Corriegills bed, hard, splintery, and breaking into small prisms. It is seen in the sandstone slopes above, but does not intersect the columnar façade of greenstone overlying the sandstone—this fixes the relative age of these two erupted rocks. The sandstone now first appears on the shore, and is immediately traversed by a trap dike 4 yards wide, ranging N. 10° W., and this, as well as the claystone, produces a decided alteration on the structure of the sandstone. But there is nothing on this part of the coast that equals in interest a pavement of basalt which extends for 200 to 300 yards in breadth, and from the sea line to the base of the cliffs. This causeway forms the nearest approach we have anywhere seen to the basaltic pavements of Staffa and the Giant's Causeway. The rock is not all a pure basalt; much of it has a fine porphyritic structure. The pillars are less perfect, less various in form, and are without the cup and ball articulation, the test of a perfect pillar; and many cracks crossing the surface somewhat mar the uniformity of the prismatic structure. This latter defect is perhaps owing to the action of the tide, which covers the greater part of the bed; a portion of which, indeed, seems to have been already removed by this cause. As this great lava stream rushed through the sandstone strata, pieces were torn off, borne along, permanently entangled in it, and are now found isolated in the basalt and altered to a crystalline structure. But there was another later irruption—the basaltic bed is traversed by a greenstone dike 5 yards wide, and ranging W.N.W. This, as well as other dikes, traverse to the very summit of the cliffs, intersecting the overlying columnar greenstone, which is thus older than the greenstone of the dikes, but newer than the claystone already noticed as clearly erupted. We have thus here erupted rocks of three ages.

80. The sandstone, gradually ascending southwards, now rises to the edge of the cliff, and the overlying trap retires inland, but continues to occupy higher terraces over the sandstone. In front, near the cottages, the shore line is low, and from this point southwards the trap appears in many interesting relations with the sandstone, to which we can only invite the observer's attention, without attempting to describe them. The trap is an injected bed, and occurs in some places overlying, in others in repeated alternation with, the sandstone, and again the two rocks come together at the same level. The dikes, too, are very singular; they make strange turns and bifurcations, and one consists of alternate bands of trap and sandstone. Presently after, a low cliff begins and runs westwards by Kildonan Castle. This interesting old keep stands

on the front of a flat terrace, which runs back northwards to the trap hills, and faces the sea in a cliff 14 or 15 feet high. Its foundations are laid on a rock of columnar clinkstone with much iron, like that of Holy Isle, and which changes inland to what is conveniently but incorrectly called a trap porphyry, a gray or blue compound base with earthy felspar imbedded.

81. Kildonan Castle is a square keep, without ornament, four

storeys in height, with several vaulted apartments. The date of its erection is uncertain; but it was probably put up at the time of the war with the Edwards, as one of a line of watch-towers reaching from Ailsa Craig to Dunbarton rock. Like that of Loch Ranza, it was a royal castle till 1405, when the lands and castle were granted to Stewart of Ardgowan, in whose family they remained for 150 years. To attach a turbulent chief of Cantire and the Isles—Macdonald of Dunivaig—the Earl of Arran, then Regent, forced a sale upon the Stewarts; but a few years afterwards he bestowed on Macdonald, in exchange, valuable lands at Saddell in Cantire, and became himself owner of Kildonan. The Stewarts, of two families, seem next to have got the property; first, one under Cromwell, the Hamilton estates having been forfeited; and again, another by succession. They went afterwards to the Earls of Bute, from whom they were purchased, about thirty years ago, by the Hamilton family. The late Duke, when Marquis of Douglas, erected the handsome shooting-lodge which now stands on the summit of the Dippen cliffs, with a magnificent look-out over rock and flood.—Kildonan Inn is by the shore, under the platform, where we may rest for a little and partake of its hospitality.

82. The claystone platform on which Kildonan Castle stands extends inland as a bed in the sandstone, which also shews below it on the shore, and forms the basement of the terraces and ridges of trap which sweep round westwards by Benan and Lag. This trap is mixed up with claystones, and dark-based felspar and pearl-stone porphyries, in such a way that it is impossible to unravel the order among these products, if order there be. Sandstones, shales, and marls occupy the shore, and stretch far up the river-courses and lower parts of the glens, whose higher sides exhibit in section the series of traps and felstones just alluded to. It would be desirable to have all the strata thoroughly examined for fossils, as these, if found, would fix the age of the beds. These interior sections are more deserving of inspection than the shore, which shews

only a repetition of appearances with which we are familiar—dikes of greenstone, claystone, and rarely pitchstone, some of the dikes standing up like walls, and numerous granite boulders of moderate size. There are in all about forty dikes, and the range is generally a little W. of N. One of the largest is that near Kildonan, forming the little harbour of Drum-la-borra. A dike attaches Pladda islet to the mainland, and has suffered so little disintegration in this exposed situation that at high water a vessel can cross only at one place.

Auchinhew streamlet comes down opposite to Pladda, and shews some fine sections, and a very singular and highly-romantic waterfall and pool called Eais-a-Mhor, or the great fall. The fall is N.E. of the Auchinhew farm-house, N. of the road; but the section lower down should first be visited. There are here several alternations of columnar trap and amorphous trap, with sandstone and marl. The horizontal joints and vertical seams are both very distinct, and mark off most of the pillars as pentagonal prisms. At the fall the water descends through a height of 70 or 80 feet into a large circular basin, or amphitheatre, the perpendicular sides of which present a façade of pillars resting on sandstone. Over this is another bed of sandstone, surmounted by beds of amorphous trap, which forms the channel of the burn above, towards the moors, and by its hardness has prevented disintegration farther up. The front of the fall is formed by a dike, which, when it reaches the basaltic stratum, divides into two branches and terminates, thus clearly shewing that it formed the vent of eruption. The sandstone is indurated along the lines of contact, and the strata shifted out of parallelism. The effects of so small a stream are quite wonderful to contemplate here, and suggest the idea that it must have been of larger volume in bygone times. Ballymenoch stream, a little way east, shews somewhat similar appearances; and farther west, near Benan-head, on the north side of the road, there is a fine section of the same kind. The stream has cut down the strata to a depth of 70 feet, in a fine cascade; and a trap dike, after traversing the sandstone strata, loses itself in the overlying greenstone—another clear case of a lava-chimney. Calcareous nodules, rarely exceeding six inches in diameter, are disseminated through the sandstone, both here and in several other sections; and Mr Headrick mentions (p. 117) that a small cascade, on a rocky front south of Auchinhew burn, has deposited a mound or small hill of stalagmites, or lime incrustations, shewing that lime prevails in all the sandstones here, and hence that these probably are the true carboniferous beds. The high grounds north of Benan-head, which reach 523 feet in altitude, consist of felspar porphyry, and greenstone, the former generally in a superior position, but in some places coming against the latter at the same level.

83. The Struey rocks and Benan-head form, both for the geologist and botanist, one of the most interesting features of this coast. The headland is 457 feet high, and is a mass, from top to bottom, of igneous products, variously intermingled, but consisting chiefly of a dark-based felspar porphyry, basalt and greenstone. The crystals imbedded in the dark, compact felspar base are of that variety of felspar called pearlstone, with which are commingled quartz crystals of a smoky hue. In the greenstone there are dark-coloured crystals of hypersthene with the lustre of crystals of oxide of iron. It is difficult to make out exactly what are the relations of these rocks to one another in the headland. In some places they are intermixed, and perhaps alternate in bands; but the porphyry, as a mass, may be said to prevail upwards, the two other rocks being in greater proportion below, except that on the east side the porphyry comes down to the level of the beach. The entire mass of the headland is wedge-shaped, narrowing downwards; the upper edge of the porphyry cliff being on the same level on either side as the strata of sandstone, the ends of which abut against the trap rock throughout. Though the sides converge rapidly, the termination of this wedge-shaped mass is not seen, the igneous rocks running out

under low water and cutting off the sandstone. Dikes of common trap intersect the whole series from top to bottom of the cliffs; and a little way west of the headland a dike of quartziferous porphyry, very distinct from the porphyry of the headland, intersects the sandstone on the beach.

The Black Cave in the front of the headland has been excavated in a vein of rotten trap, between two basaltic walls, and extends about 40 paces by 14, with a height of 70 or 80 feet. Rocks have fallen from the roof, making an opening overhead. The vista across the sea is finely closed by the erupted cone of Ailsa. Loitering here may turn out to be very embarrassing, as the front of the cliff is not passable at or near high water; and at certain times of tide we may get into such an awkward situation that our only chance of escape is "to swim for it."

The cliffs westwards present for some distance porphyry and greenstone, apparently in alternating bands; but it is difficult to make out the exact relations. Thence to the mouth of Torlin water, the shore and cliffs are occupied by variously coloured sandstones and marls, with occasional nodules of limestone and ironstone. The platform of the beach is crossed by multitudes of dikes, mostly of basalt and greenstone, which usually produce a decided alteration on the sandstone strata. The road from Benan to Lag, which runs a little way inland, shews generally, along the undulating grounds on either side, beds of claystone beneath the surface soil; in the hollows in some places greenstones appear, either the tops of dikes shewing here in their inland course, or the edges of beds emerging from under the claystone.—After our hard day's work, we are glad to find that Lag is less than three miles distant from Benan-head. At a turn of the road and fall in the ground near the Kilmorie post-office, we see the old church of Kilmorie, perched on a high bank over Torlin water; and soon the welcome sight of the peaceful hollow and neat inn greets the eye—we shall find few inns anywhere in which there is the same attention to our wants, and the *menage* so unexceptionable.

EXCURSION XV.

LAG TO THE SHELL-BEDS.

84. LAG is an admirable centre for several excursions. A thorough examination of the shell-beds, especially with the purpose of collecting the leading species, will take two days at least. There is much to see, very curious and interesting, by following up to their sources the two largest streams, Torlin and Slaodridh waters; Benan-head and Dippen-cliffs may very well be visited from it, reversing our order; as also the shore both ways, east to Benan-head, and west to the porphyry of the Brown hills. For the Botanist and the Antiquary there is a fine field; and to a lover of the "gentle art" the two largest streams of the island, with their two considerable tributaries, will afford sport for several days. But "the Lag" has other charms; its situation is peculiar and romantic; its complete seclusion and perfect quiet wonderfully tranquillising; though warm and sheltered in its sunny dell, the inn is not ten minutes' walk from the sea—an early stroll there braces for the day's work; the country about is the best cultivated and richest in Arran; and it is delightful to saunter at even-tide among the sweet, teeming fields, to mark the fitful changes of the sea, to watch the sun go down behind the Cantire hills, and catch his last rays gilding the dome of Ailsa.* "The Lag," as it is called in the neighbourhood—that is, the Hollow—owes its peculiar features to a speciality in the rock formations:—there has been an upthrow and fracture of the sandstone-beds, here soft and marly, attended by the appearance of great dikes of basalt; the stream, issuing by a narrow opening

* The farms of Clavich (Mr Robert Speirs) and Torlin (Mr Finlay Cook) are noted for the admirable methods pursued upon them, and the successful results.

and liable to great floods, would readily disintegrate and sweep off the materials of these fractured beds. The part most deeply imbayed is that immediately opposite the entrance of the stream and its tributary into the existing hollow, below Kilmorie Church; from this the sweeping currents would be reflected to the heights north of the inn, and thence again towards the point where the stream escapes. The principal dikes are at this point, below the bridge. The largest is 30 to 40 feet wide, ranges nearly N., and forms the high cliff over the flower-garden; another, farther S., is 10 feet broad, and ranges nearly N.E.; a third, 9 to 12 feet wide, ranges N. 10° E., and dips W. at 80°. Above the bridge there are two large dikes; the lesser stands high above the south bank of the stream, like the side of a former dam, and is seen on the north side in contact with disturbed and altered sandstone—width 25 to 30 feet, range N. 26° W., dip E. at 70°; the greater, farther up the stream, is 35 to 40 feet wide, and ranges N. 5° E. The sandstone strata in the hollow dip usually about S. 10° to 20° W., at angles varying from 20° to 30°; but there is considerable disturbance. The heights round the basin vary from 50 to 70 feet above the stream.

85. The beds of arctic shells most accessible from Lag, and the most satisfactory to examine, are about three-quarters of a mile distant, and on the banks of a tributary to Torlin water, called the Cloinid or Clenid burn. It joins the Torlin on its right or west bank, a little way above Lag Inn, coming down from the northwest through a deep rocky gorge, easily passable when the water is low. Another way to reach them is to follow the high road as far as a lane near the Kilmorie school-house, pass by this to the river, over which there is a wooden foot-way, and then to cross the fields, by a cottage on the high bank opposite the church, down into the channel of the Cloinid burn. We are thus on the shell-beds at once. They rise steeply to a great height on the east bank of the stream, a very little way to the north of the upper end of the gorge. Fragments of shells on the surface soon shew that we are in the

right place. In an excursion to these beds, and the others in the neighbourhood, running through three days, I was accompanied by the Rev. H. W. Crosskey of Glasgow, who has carefully studied the shelly deposits of the Clyde basin, and east of Scotland; and the result of our joint observations is the account that follows:—The object of enlarging the list of species already given (Art. 21) was quite subordinated to that of working out the physical order of the beds, and establishing among them definite relations. It has been already remarked (Art. 21) that under the term boulder-clay or till many distinct beds have been classed as one formation; whereas there can be no doubt that if the subject were properly investigated and understood, order would reign among these as among the other strata of the earth's crust, and that beds of a like mineral character would have also a definite stratigraphical position. Not that such a series of beds is to be supposed to occur universally—for this is not the case with the solid rock formations themselves; but that if there be a considerable development of these "superficial beds" in different places, they shall be capable of subdivision into several terms with distinct and persistent characters, and definite relations to one another.

Now, in attempting to establish a physical order among these beds, we are met by two difficulties which we do not encounter in the case of rocky strata. The upper layers under the soil are loosely aggregated and pervaded by fissures which admit water, while the lower layers are retentive of water. Thus a great hydrostatic pressure comes to be exercised, and as the beds have not that continuity and coherence which enables one part to lend support to another, the upper beds slip down upon the lower. Such landslips also occur by the undermining of a stream. In this way the order of the beds is completely masked; a bed 10, 20, 30 feet higher in the series is placed in front of, perhaps, the lowest bed, and so remains for a long period, till by successive slips or underminings the whole fallen mass is removed, and the natural order of the beds is again revealed. In most cases satisfactory evidence that such landslips have taken place will be had by examining the grassy brows along the summit of the cliffs. Then again, the beating of rain, the action of the air and sun, and the trickling of runlets of water over the front of such a section, produce a sort of general *wash-over*, as we may call it, which gives the same facing to all the beds, and completely obscures the natural succession. There is a third obstacle and cause of error, like what one sometimes meets with in the case of the rocky strata—the difficulty of finding junctions. The wash renders it impossible to see the line of contact, and it is often necessary to search over a large surface. The observer takes the field armed with a hammer only, but finds his weapon, good for other purposes, of no use in this contest. A pickaxe and spade are required for opening up the beds, and clearing away loose matter, in order to obtain a distinct view of the planes of contact.

The difficulty of finding a true section, and so determining the actual succession of the beds, is now apparent; and the great probability, that, under the name of boulder-clay, beds really distinct have been included. And we cannot help thinking that much of the confusion which prevails has arisen from overlooking the causes of error now adverted to. One of the sections under consideration was visited by us three times, and another twice before we were able to satisfy ourselves about the order of the beds. In several places we left the sections completely opened up for the benefit of future observers; it will probably be many months before they are obliterated by the causes already alluded to.

86. The base of the section in the Cloinid burn is the true old boulder-clay resting on the upturned edges of the sandstone rock, which dips nearly S.W. The depth is about 15 feet in the lower part, but the upper surface declines southwards, or down the burn, so that the bed becomes thicker upwards. The clay is excessively hard and tough, and the hammer makes no impression upon it; the imbedded stones

are of all sizes, from small pebbles to large boulders, all confusedly mixed, and many striated, smoothed, or polished. It is very striking to observe the prevalent rounding-off upon the edges of a large proportion of the stones, and the high polish which has been given to them, and *that* upon all alike, of whatever size. This seems clearly to indicate a powerful agent acting for a long period. It cannot have been a continued transport in water, for, as already observed (Art. 22), this action would have left no trace of striation. The majority are local, but comparatively few of sandstone; porphyry, greenstone, and syenite, which form the higher grounds and the hills northwards, abound, but there is also granite of the coarse-grained or Goatfell variety, and slate, both, of course, from the northern mountains. The ordinary wash covers, in most places, the surface of the boulder-clay, and in this wash adhering to the surface, we found a few small fragments of shells; no fragment in the clay itself.—Over this is a clay bed, the repository of the shells, sinking southwards with the upper surface of the bed below. It is 7 to 8 feet thick, and is a hard, compact, rather fine and dark clay, with a good many small stones, not striated as a whole; our experience, indeed, is, that few were striated. The deposit is very different from that below, much less compact, more easily worked, and the contrast of the stones is remarkable. The upper part of the bed is of a reddish colour, and a little sandy, while, in the lower part, the colour has more of a bluish tinge, like the boulder-clay, shewing that it was formed on the spot, and partly out of the boulder-clay, on the bottom where the shells were living. The arctic character of this bed is clearly shewn by two of the species found in it—the *Pecten Islandicus*, and *Astarte borealis*—the only two absolutely arctic. By far the greater number of shells are in single valves or in a broken state, yet not so small but that the species can be determined. This fragmentary condition of the shells shews a very disturbed state of the waters along the coast on which they lived. But even on stormy sea-coasts, with stony shore, sandy mud, and broken shells, there are sheltered places inside of points, and behind banks and ledges of rock; and there can be no doubt that, somewhere about here, if one could happily alight upon it, there must be some such bed of unbroken shells indicating a quiet deposit.

The following section gives the succession of beds at this place in ascending order:—

1. Boulder-clay, 12 to 20 feet thick.
2. Clay-bed with shells, 7 to 10 feet.
3. Dark sandy bed with open texture, 4 or 5 feet—apparently local.
4. An upper drift of sand and stones; thickness variable.
5. A compact bed of stones, with less sand, forming a marked line on the cliff, but not easily reached; thickness 5 or 6 feet.
6. An upper drift, similar to No. 4.
7. Surface soil.

These upper drifts, covered in front with their facing, or wash, look very like the boulder-clay; but a closer examination, and removal of the false face, shew them to be very different. They are much less compact, more loose and easily broken, generally more sandy; and the stones are mostly angular, or, if polished, the surface has a finish very inferior to that of the stones in the boulder-clay; as if they had been exposed, since the polish was put on, to some continued wearing action. In fact, they have very much the appearance of a river deposit; and in looking at some of the heaps of detritus left by floods on the sides of the stream below, it struck us how very like they were to these upper drift beds. The young observer could not employ a day given to geology to more purpose than in examining these upper drifts, and contrasting them with the boulder-clay, in some typical section.

The shell-bed No. 2 is seen over the boulder-clay, from the extreme southern end of the section at the steep bank over the stream, across the whole front of the section, ascending, however, northwards; and the same species are got in it throughout, there being a great many fragments of *Pecten Islandicus*, *Cyprina Islandica*, *Modiola modiolus*, *Astarte borealis*, *A. compressa*, *A. elliptica*, and occasionally single perfect valves. The part of the bed most accessible for working is on a slip towards the southern part of the section.

A little way up the stream, on the same side, there is a very high bank with apparently a like succession of beds, but so steep that it is very difficult to get a footing; and, as the day was now far spent, we were obliged to leave them unexamined.

87. On the west or right bank of Torlin water, opposite Kilmorie Church, and a little below the wooden foot-way, the shell-bed is again seen. The sandstone strata here dip N.E., opposite to the usual direction; the stream for a short distance runs upon a whin dike, and there are beds of trap farther down on the opposite bank. The shell-bed rests on the sandstone, thinning out southwards in the bank. The section has a small horizontal extent, the bank northwards being broken and covered with wild shrubs; but there are indications of the existence of the boulder-clay in this direction, and it probably comes on here under the shell-bed as the ground rises up the river. Over the shell-bed there is a red sandy bed, 12 to 15 inches thick, over which there is an upper drift of earth and stones. The shell-bed has the same character as in Cloinid burn; it is tough and compact, dark-coloured, with small stones, and the same species of shells. Not having previously compared the two beds in Cloinid burn, an observer would be very apt to confound this shell-bed with the boulder-clay.

88. We pass now to Slaodridh-water, about a mile west of Lag. Less than half-a-mile above the bridge, a large tributary from the north-west joins the Slaodridh. This is the Crookcrever-burn, on whose east bank in several places the shell-bed occurs. A very little way up the burn there begins a highly picturesque gorge, with steep banks of wild wood, cut

by the stream through beds of variegated sandstone, alternately hard and soft, crumbling shale falling fast away, and the hard sandstone projecting in ledges. A good way up, this gorge expands into a wide open space, and high clay banks form the eastern side, but immense slips greatly confuse the order of the beds. After a repeated and careful survey, and having eliminated the various sources of error, we made out the succession of the beds as follows:—Sandstone rock is the base of the whole section; over this a bed of the true boulder-clay, 30 or 40 feet deep, reaching about half-way up the bank, the clay being of the usual hard, compact, almost unworkable structure, and with some very fine examples of striated stones, blocks, as well as others of lesser bulk. Its upper part becomes, through a space of nearly a foot, a hard, dark-coloured, gravelly sand, extremely compact and obdurate under the pick-axe. On the top of this, and strongly contrasted with it, rests the shell-bed, of almost exactly the same composition as that already described in the other two sections; a dark compact clay, with small stones rarely striated—very different from the boulder-clay in the comparative facility with which it is worked. In addition to the shells already named, most of which were found, there was a perfect *Leda*, the valves united, and fragments of *Balanus.* Above this bed are the upper drifts, very similar to those already noticed, and here, as in the former case, very strongly contrasted, if carefully examined, with the boulder-clay; yet there can be very little doubt that they have often been confounded. This section shews the series very distinctly, and it has been left by us completely opened up.

89. The last section opened was just below the east end of the sandstone gorge, opposite a bend on the river, and a little way above a wooden guard placed upon the stream. Here a blue argillaceous shale, dipping S.W. at 15° to 20°, forms the bed of the stream, and rises up with the same dip into the east bank. Over this the boulder-clay and other beds come on; and, following apparently the inclination of the shale be-

low, they slope towards the stream, and so thin out westwards. This arrangement is shown in the annexed cut:—

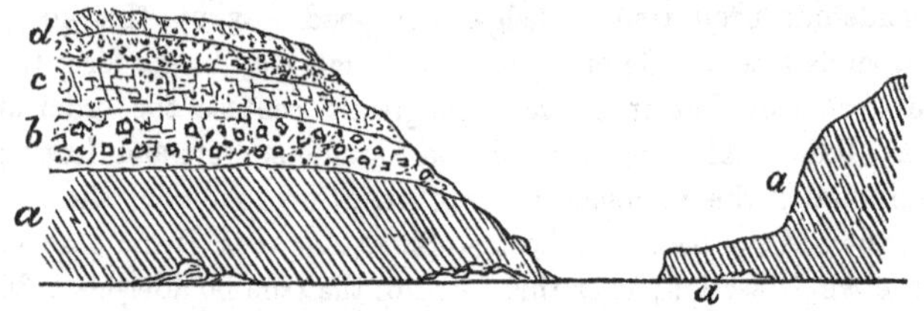

(a a a) shale; (b) boulder-clay; (c) shell-bed; (d) upper drifts.

At the expense of some time and labour, the facings and debris were here completely cleared away, so that the entire section was laid bare, from the underlying shale to the upper drift bed; and nothing could be more interesting or perfectly satisfactory than to observe the contacts and complete distinction between the contiguous beds. The boulder-clay has all its usual characters, as already described, and does not yield a single shelly fragment; the shells begin to abound directly the overlying clay bed is entered. This has its usual character, which we need not again describe, and over it come the upper drifts, again strongly contrasting with it and with the boulder-clay, but here not well exhibited, from the nature of the ground.

90. This section is highly illustrative in another way. We often find that, either from their successive erosion, or from some inequality of original deposition, these beds thin off at the out-crop, just as rocky strata do; and thus the beds are brought so close together that the distinction is not perceived by an unpractised observer; and that shells from an upper bed get on to the surface of a lower, or perhaps into its very substance. This source of error is most likely to be met with in the case of basin-shaped or nearly level deposits. We think it most important that all who engage in the study of these most perplexing superficial formations should guard against this source of error and those already indicated (Art. 85); they ought first to study the mineral character, and then the order

of the beds, and not hastily assign to the boulder-clay—that is, the lowest and oldest bed—shells or other fossils which may really belong to those which are superior to it. We have never found fossils in or under the boulder-clay, but we by no means assert that these do not exist in it and under it; yet we cannot help thinking that far greater caution should be used in assigning fossils to this bed than has hitherto been shewn. We cannot theorise with any certainty respecting the origin of these deposits until, by a more careful study and extended examination, we get hold of a number of great general facts in regard to the physical order of the beds, their internal structure, and fossil contents. In regard to the fauna of the Arran beds, it will have been observed that it is all of one character; the same species occur in all the beds, with slight local exceptions. The shell-bed itself is of uniform character, with very slight variations; and hence it follows that uniform conditions affecting the marine creatures prevailed over this district. But the area is too limited to warrant any general conclusion. The probable conditions have been already noticed (Arts. 20–22). The boulder-clay may have been formed by land ice; the shell-bed over it, under and around a rim of ice when the land had been depressed; and the upper drift beds partly by local glaciers as the ice was disappearing, and partly perhaps by rivers, then of larger volume than now, as already hinted. The elevation of the shell-bed, or of the upper surface of the boulder-clay (Art. 21, last sentence, in which this idea is obscurely expressed), taken in connection with the depth due to the species, gives us the measure of the depression of the land. Now, the elevation of the beds in question ranges from 70 to 180 feet; so that the greatest depression of which we have any evidence here is 180 feet below the present sea level, not taking account of the depth required for certain species; and we hesitate to speak with confidence of any greater depression, with our present knowledge of the beds farther up the glens. In Clachan Glen there are beds of at least equal altitude and enormous thickness; but their altitude has not yet been determined.

91. The bed and banks of these four streams present many objects of interest besides the shell-beds. They exhibit sandstones and shales with nodular limestone, and sometimes nodular ironstone, among which no fossils have yet been found, and these rocks are pervaded by traps and felstones of all varieties, both in dikes and overlying masses. Near the head of Crook-crever burn claystone appears in both relations: on the banks it overlies, in the bed of the stream it breaks through the sandstone in conformable beds, and produces a very decided change upon it, altering it to the condition of a fine white quartzite, a very clear proof that the claystones are not altered sandstones, but themselves erupted rocks. The bed of the Slaodridh at Glen Rie mill, and the country thence to Lag, and again far up the stream on the east side, show felstones as the prevailing rock—on the west side from Glen Rie upwards the prevailing rock is greenstone. At Glen Rie mill the sandstone is greatly disturbed and altered by a prismatic compact felspar, with bits of quartz and rarely felspar, which forms a bed in the sandstone. We might return from Lag this way to Lamlash, and trace the carboniferous strata across the watershed of the Ross. We might pass also up the banks of the Torlin, and trace out a wondrous variety of curious relations among the traps and felstones which abound there, visit a Stone Circle not far from Urie Loch, examine the syenites and pitchstone vein which break out on Urie hill, and then pass down to Whiting Bay or Lamlash. An account of the rocks to be seen on both these routes would lead to great detail, and imply much repetition; but the walk of ten or eleven miles will be found very instructive.

On a bank near the sea, within the first field east of the mouth of Torlin water, there is a fine specimen of the sepulchral cairn. It consists of a great number of stone chambers formed by five large slabs, said to be filled with human bones. One only has been opened, and from it certainly bones and a skull were removed. The lid of the stone cist or chamber still lies near; an upright stone probably marks the head. It was once surrounded by a wall; but though now unprotected it will not be further disturbed without express authority. Nothing is known regarding its history. A little to the west of the mouth of the Slaodridh, on a high, steep mound near the shore at Haddock Port, there are the remains of a fort called Tor-chastel, or the castle hill. It is connected to the mainland behind by a narrow neck, which is guarded by an outwork with a stone foundation. On the top there is the circular foundation of a Danish fort, consisting of a wall of large stones without cement, 5 feet thick, and having a circuit of about 52 yards. It was doubtless made strong to defend this place of landing for enemies from Cantire, Haddock Port being the extreme south-western point of the island, and the only port here. Bones of man, the boar, deer, and ox have been found among the ruins; and the surface has been in later times cultivated, as ridges are marked on the summit.* Not far from this point is the natural harbour of South End between the mouths of Slaodridh and Torlin waters. This is formed by four dikes, of which two run nearly N. and S., and two nearly E. and W.; of the two latter, the seaward one is highly prismatic, and stands up like a wall, but is breached, and admits vessels, while it forms a shelter for them. The other forms a rude landing pier on the north side.

The old parish church, called St Mary's cell, or Kilmorie, existed early in the fourteenth century, about the time of Bruce. It was however a mere cell, being only 19 feet by 10. The present church, built in 1785, is near the site of the old chapel, and is said to have the font which belonged to it.

* M'Arthur's Arran, p. 80.

EXCURSION XVI.

TO KING'S COVE AND THE STONE CIRCLES.

92. To King's Cove is a favourite excursion, and deservedly so, for the scenery there and about Drumadoon is extremely beautiful. With this may be combined a visit to the new granite tract of Craig-Dhu, the wonderful assemblage of pitchstone veins, and splendid group of stone circles, at Tormore. In Arran, as in Switzerland, the success of a day's work depends upon an early start. But we must be sure of the day; let it be calm and sunny, if the scenery on the charming Drumadoon shore is to be enjoyed to the full. Leaving Invercloy, we turn to the left, pass Brodick Church, and begin the long ascent of the String Road. On the right we look down into the "Sleepy Hollow" of Glen Shirag, and note with surprise what work of excavation the little streams have done at their last joyous leap into its depths. On our left the ground swells rapidly up from Brodick wood, and culminates in the high porphyry ridge of Windmill Hill. From behind the smooth ridge of Old Red in the foreground, the granite peaks and crests rise up in succession as we ascend, and appear in new and grand aspects. As we breast the slope, we turn often to gaze in wonder and delight on the lovely scene we have left behind—the smiling plain of Brodick, the glassy bay, and hanging castle woods. When we reach the summit there is before us a fine vista, shut in by two long declining ridges, and closed westwards by Kilbrannan Sound and the hills of Cantire.

After the watershed is passed, granite appears on the left hand—the north-west limits of the Ploverfield district—and runs a long way down the side of the hill; then bands of syenites, the same as those which succeed the granite on its west border (Art. 11), strike in upon the road, intermingled with sandstone, here of carboniferous age. The bottom of the glen and steep ridge on the right are composed of the old red sandstone. The old bridle path runs at the base of this ridge. On its smooth outline it is curious to notice watercourses and glens in an absolutely initial stage; such is the force due to the height of descent, that water collected into the merest rill, by the slightest possible inclination, has a great power of excavation, and speedily forms a perceptible channel and considerable talus. Leaving the road at the Glen Luigh bridge, a bed of limestone will be found high on the hill-side to the south, by following a bridle road leading up to the old quarry. It is a crystalline lime without fossils, much altered by trap veins which traverse it, and to which the limestone adheres firmly. It seems to be near the base of the carboniferous series, in the same position as the limestone at Achab farm and the Fallen Rocks.

We are now close upon the Craig-Dhu granite, already fully described (Art. 12). Craig-Dhu is a high steep cliff, fronting that part of the Shiskin road where the road to Dugarry parts off from it. A talus of blocks strews its base. In July, 1855, when driving along this road with a party of friends, I noticed how unlike these were to sandstone blocks, and was thus led to the discovery of the granite tract. It struck me as remarkable that on a route so frequented by geologists, the occurrence of the granite here had so long remained unnoticed.

A little farther forward, opposite the pillar letter-box, the Dugarry road turns off to the right, crossing Mauchrie water by a substantial bridge, and passing through the stunted remains of an old forest; the distance is about three miles. The only house on this road is the farm-house of Mauchrie, opposite to which a path leads down to a wooden footway across the stream and near the shore, by which we can easily reach the Tormore pitchstones. The stone circles, a mile inland, and due S. of the farm-house, may also be reached this way by a very short route. But we shall pass up to them from the shore, after visiting King's Cove and the pitchstones.

From the wayside post-office we continue our course to Blackwater-foot, a distance from Invercloy of eleven miles, passing on the left, the glens of Craig-Dhu, Balmichael, and the Clachan, and having a wild moor on the right, on whose northern brow the tall standing-stones of the Druid Circles arrest our gaze. At two miles distance we pass Shedok Inn, and two miles farther, turning sharp to the right, we are at Blackwater-foot, the extremity of the vale of Shiskin. The unpretending little inn here will afford very comfortable quarters for the night should we find occasion to remain.

93. On the south-west coast, in both directions from this point, there occurs the greatest development in the island of the felstone family of igneous rocks—a peculiarity for which

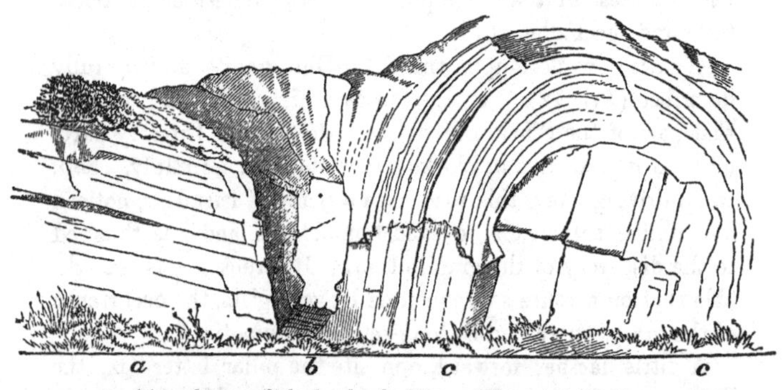

(a) sandstone; (b) broken band with cave below; (cc) claystone.

it does not seem possible to assign any cause. At Blackwater-foot a great bed of indurated prismatic claystone occupies the bed of the stream, the sides of the little port, the shore for some distance and a long line of inland cliff. Its outgoings N. are lost amid the sands. In some places it forms very perfect columns, with a near approach to the true cup and ball jointing. In the cliff it intersects the sandstone in the form of a great dike; and in one place is seen to turn up the ends of the strata abutting against it; in another place the beds approach it at an inclination, and there is a mass of broken rock between them, a small cave being formed in this at the bottom. Here, as shewn in the annexed cut, the mass of claystone on the south side is divided into great concentric curved bands, arranged round a prismatic nucleus, and exhibiting on a large scale that peculiar structure of the rock, often seen in the fracture of hand-specimens, which it has in common with the other felstones and the ordinary traps, and which comes most strongly out in decomposition. The appearances here are strongly illustrative of the eruptive character of the claystone.

The sands stretching hence to Drum-a-doon exhibit many dikes, ranging in various directions, of various widths, and consisting of basalts, greenstones, both common and of porphyritic structure, and felstones. The botany of the sands is rich; the surface has a thin carpeting with a profusion of beautiful wild flowers. On approaching the grand façade of columns forming the precipice of Drum-a-doon, we find at its south-west base on the shore a great bed, or perhaps dike, of felspar porphyry tilting up the sandstone at a high angle, in no way connected, so far as can be seen, with the overlying porphyry above, but running out into the tide-way, and forming a low cliff in the sands. This is Drum-a-doon Point. It is intersected by basaltic dikes, one of which has been hollowed out by the waves, and forms a landing port for boats. It is probable that these beds, and others on the south-east side, are connected, underneath the sandstone, with the overlying masses. The beds below cannot have been, at a former period, continuous with the porphyry of the façade, as the sandstone rises high in front between them. Yet the cliff has obviously suffered considerable disintegration along the front; an isolated, nearly entire, pillar stands out from the facade at the south-west corner, and the sandstone below is strewed with fallen columns, forming a talus to the precipice. These features are shewn in the cut annexed, in which the cliffs at King's Cove occupy the foreground.

The porphyry of the cliff is in columnar forms, the pentagon being the prevailing figure, and there are irregular flat jointings; the pillars are from 1 to 2 feet in the side, and the height 80 or 100 feet at the middle or highest part of the precipice; the entire height being about 230 feet. The porphyry has a base of compact felspar or clinkstone of a whitish or brownish grey or dark-blue colour, and imbedded crystals of glassy felspar or

pearlstone, common felspar, or earthy felspar, in a state of decomposition, and with an unctuous feel, like steatite; there are also transparent bits of quartz, easily known from felspar by not shewing any laminated structure, with the form and lustre of crystals whose angles have been worn off; the base is often dark and has cavities lined with black oxide of iron. The rock is thus a quartziferous felspar porphyry, often ironshot. Hence it closely agrees, in its mineral character, with the porphyry of Dunfion (Art. 42). Hand-specimens, shewing only the earthy imbedded felspar, have the look of amygdaloid; and there is a variety in which the felspar crystals have decomposed out and been replaced by calcareous spar, by subsequent infiltration; and another, in which quartz becomes a component part of the base, and the felspar crystals disappearing, the rock has a granitic structure, like that of Eurite (Art. 50). In some specimens the imbedded felspar has a nucleus of quartz. The pillars rest on a laminar stratum, consisting of the base of the porphyry without the felspar crystals, and thus resembling a dark claystone, metamorphic sandstone, or some varieties of the lighter coloured basalts. A thin stratum of white sandstone is interposed between this bed and the underlying red sandstone. Two dikes of greenstone traverse the cliff right up through the columns of porphyry. There is a path by the base of the precipice, which is about 500 yards in extent; and off its N. W. termination there are masses of a similar porphyry on the shore and in the low cliff, whose relations to that of the great façade cannot be made out.

Drum-a-dun, the hill with the fort—Drum or Drim, whence the Latin *Dorsum*, is a rounded ridge—is so named from a fortress on the summit, which seems from its position and remains to have been one of the strongest places in Arran in the olden time. A wall 8 or 10 feet thick, founded on huge blocks, starting at either end from the front of the cliff, sweeps round landwards so as to enclose several acres. Mr Headrick, writing in 1807, says—"In the middle of the arch there is a gateway, on each side of which are great heaps of stones, which seem to have been additional works for its defence; within the enclosure are several ruins of houses of loose stone. . . . Had it a sufficient supply of water, it might be rendered impregnable." Many of these stones have since been used in the erection of fences and cottages, and there are no traces of any ruins. The massive blocks which still remain are too numerous to have been all found upon the summit in the manner usual with boulders, and must have been dragged up the eastern slope by some powerful mechanical appliances. The great area of the place shews that its purpose was rather as a place of refuge, in case of danger, for the families and cattle of the neighbourhood, than a fort to be defended by warriors. In later times it became by universal consent a sanctuary or place of refuge "from the assaults of enemies, for whatever number of men or cattle could get within it." *

94. The coast now sweeps round in a fine bay, whose grassy floor and low cliff of red and white sandstone contrast finely with the rugged features of Drum-a-doon. Northwards among these sandstones there are inclined beds of green pitchstone, which alter the sandstone strata in contact; their termination cannot be traced. There are also bands of claystone in the cliffs. Rounding the angle of the bay, we come on high cliffs of soft yellow sandstone, deeply excavated by the sea, when it stood at the higher level so often already alluded to—now, and for more than 2,000 years, above that ever reached by the highest tides. To one of these great interest has been given by the tradition, that it was for a short time the residence of King Robert Bruce, on his landing from Rathlin in 1307, before the seizure of Brodick Castle and descent upon Ayrshire, though there appears a greater probability that Loch Ranza was the port where his flotilla found shelter. A hard vein of sandstone had stood in the cliff vertically, and the strata of softer rock declined either way, from some early fracture; on this the sea working for ages scooped out the cave. The hard vein is still seen in the sharp Gothic-like arch above, in the position of a keystone, and in the back part of the cave comes down to the floor, forming a column with a recess on either side, 30 feet in length. The cave is 100 feet long, 50 wide, and 55 high. The column has figures rudely cut, representing a two-handed sword and a cross. There are also rudely-drawn plans representing a scene in the chase, and also sheep, goats, and cattle.† "I felt a holy veneration," says our worthy author, "while exploring the cavern where Bruce had sheltered; not because he was a king, but because he was a patriot, a hero, and an assertor of

* Martin's *West. Isl.*, 1759, quoted in M'Arthur's *Arran*, p. 86.
† Headrick's *Arran*, p. 160.

the independence of his country;" and then pictures his struggles and aspirations while working here his great plans, and ends by an eloquent denunciation of all warriors who have fought for other objects than those which Bruce pursued.

The great hero of the Gael, Fhion or Fingal, is said also to have resided here, and to have had a son born to him in the cave; and a straight groove is shewn on the side of the cave, more than 2 feet long, said to be an impression of the child's foot the day after his birth; from which Mr Headrick, with great gravity, goes on to calculate that Fingal must have been 70 or 80 feet in height, and his wife 50 or 60 !

There are several other caves near the King's Cove; a large one on the south side, with two entrances and a huge pillar between, is "the King's stable;" smaller caves on the north are "the King's kitchen, larder," and so on.

95. The King's hill, "Tor-an-righ," a very little way east of this, is the highest point on the shore here; and has probably given the name Tormore, or the great hill, to the townland. On its summit there are many granite boulders of the fine variety, among which one was distinctly marked with glacial striæ. The cliff here is the side of an immense dike of felspar porphyry ranging nearly N. and S., and fronting the sea in a mural precipice 60 or 70 feet high, along the base of which there is no passage except near low water. The porphyry is very like that of Drum-a-doon, and the dike is 80 or 90 feet wide. Placed longitudinally *in this dike*, westward of its middle part, there is a greenstone dike about four feet wide, running out with it seaward to the low point where the porphyry dips into the water; it is also crossed *diagonally* by a greenstone dike, which bifurcates upon it, and is seen far up on the hill-side above. Alongside the dike there is a large cave. The finest set of dikes to be seen in Arran occurs here, exhibiting, in a small space, all the members of the igneous series, greenstone and basalt, porphyry and trap porphyry, pitchstone, claystone, and hornstone. To understand the relations of the various dikes, the shore should first be traversed several times, the

whole extent being only a few hundred yards. The dikes range over the sandstone platform to the north-west of this. The principal dike, traceable continuously for a long distance, is formed of green pitchstone; it rises from the sea southward with a range 40° E. of N., a width of ten or twelve feet, and a S.E. inclination of nearly 30°; but the course undulates 35° or 45°, bending towards the west, or into parallelism with the shore, towards its northern termination. There are veins of slaty hornstone on both sides, next the sandstone, and on one side a thin layer of basalt. On the side of the great vein next the sea, numerous veins enter it nearly at right angles. First, a vein of hornstone six to eight inches wide; next, one of basalt five feet wide. Another vein running a little N. of W. has on its north side three feet of pitchstone, passing into hornstone, then four to six inches of claystone, and on the south side fifteen feet of basalt or fine greenstone. Next is one thirty feet wide running 35° W. of N., consisting of basalt four feet, claystone fifteen feet, basalt again six feet, containing a pitchstone vein along its middle part; lastly, a vein of greenstone five feet wide. There is here also a vein of trap porphyry four feet wide; this and the others intersect the pitchstone vein, but the intersection is obscured by debris. The last vein northwards is one of pitchstone with hornstone and jaspery quartzite, running oblique to the shore, and varying from eleven or twelve to thirty feet. Just outside the tideway it meets and enters the great pitchstone vein, which, in its sweep north, has reached this point, ranging now 10° E. of N., and at the junction there is nothing visible but claystone, which forms the flooring at the common point of union. The two veins unite here; and the joint vein continues its course under masses of sandstone, but is not seen in the cliff aloft, nor can it exist there. Still farther north, near where the cliff ends, we noticed a pitchstone vein running E.N.E., seven to ten feet wide, and visible in the cliff above; and this appears to be the same vein continued. The close association of so many igneous rocks is extremely curious; we may either suppose the various products to be cotemporaneous, so that different rates of cooling and varying chemical composition may have produced the differences now observable; or we may ascribe them to successive eruptions taking place under varying conditions. Professor Jameson was the first to describe this interesting locality, which he does with great fulness and accuracy. He gives also a figure illustrating the position of the secondary or cross veins (*Min. of Scot. Isl.*, vol. i., pp. 17 and 102).

The following analyses of the claystone and pitchstone have been made by Mr Magnus M. Tait, F.C.S., assistant to Dr Anderson, Professor of Chemistry, Glasgow University:—

Claystone.

Silica,	72·50
Alumina,	11·53
Potash,	5·24
Soda,	3·37
Magnesia,	2·72
Oxide of iron,	2·06
Lime,	1·79
Water,	0·70
Loss,	0·09
	100·00

Pitchstone.

Silica,	66·03
Alumina,	12·55
Soda,	5·02
Potash,	4·13
Water and organic matter,	4·20
Magnesia,	2·33
Lime,	2·80
Oxide of iron,	2·75
Loss,	0·19
	100·00

96. The stone circles of Tormore may be very conveniently visited from this point; they lie nearly a mile inland N. E. of

King's-hill: but the readiest way to reach them is to pass northwards along the shore, after examining the pitchstones, as far as the point where the cliff ceases, and then strike inland, due E. A little beyond the farm-house of Tormore a large circle of granite blocks is first passed; a little farther, two upright stones are seen upon the moor, and when the eastern crest is reached, the group of circles is seen on its eastern slope. The tall, upright stones look majestic and solemn, standing solitary on the wild heath. We associate them with an age and the life of a people long passed away, but who were moved by feelings and passions like our own, and they thus awaken in our minds a strong human interest. The mystery of their origin and purpose enhances the charm that hangs about them; curiosity and speculation are excited: and we conceive a strong desire to discover for what purpose they were reared, and to what uses they were applied. After a careful survey of them, in July, 1860, and an examination of all that had been written about them, I was not able to satisfy myself that there was any authentic record of the discovery of relics which could throw light on their origin or use; while I was acquainted with only one or two cases in which such circles had been anywhere opened. Further investigation into a subject confessedly so obscure seeming thus highly desirable, I placed the nature and importance of the inquiry before the Duke of Hamilton, and requested permission to make a series of excavations within and around the stone circles and monoliths of Arran. His Grace not only at once most kindly acceded to the request, but expressed a wish that the operations should be conducted at his expense. His Grace also placed me in communication with his agent in Arran, Mr Paterson of Whitehouse, Lamlash, whose judicious arrangements mainly contributed to the success and rapid conduct of the operations.* But before giving an account of the excavations, it will be necessary to describe the circles in their present state. Mr Paterson was so good as to have

* Mr Archibald, land overseer at Brodick, also lent his valuable aid.

an actual survey made of them, by which their relative positions are accurately laid down by the compass, and the respective distances apart, to a scale. These are shewn in the annexed plan, on which also I have indicated the kind of circle, whether of pillared stones or blocks. The following account is an abstract of a paper read by the author of this work before the Society of Antiquaries in Scotland, May, 1862, and published in their Proceedings for 1863:—

The more eastern circle of the group (No. 1) is a single circle of granite blocks, having a diameter of 14 yards; two of the stones are entire, and stand about 5 feet high; the others merely protrude from the surface of the moor. The second (No. 2) is a single circle of tall sandstone slabs, of which three (W. to N.E.) are perfect and upright; but the rest of the circumference is defined by the bases of other stones remaining in the soil—the number being, in all, seven or eight. The pillars are from 12 to 14 feet in height, 3 to 4 feet broad, and 11 to 22 inches thick. Two circular slabs, 4½ feet in diameter, and 11 inches thick, plainly cut from a fallen pillar in modern times, and intended for millstones, lie inside the circle—one of them is pierced by a grooved perforation, as if for the admission of a shaft. The next circle (No. 3) is 13 yards in diameter; in the circumference nine stones are distinctly seen, but one only is upright and entire; its height is about 14 feet, and other dimensions about the same as in the pillars of No. 2. Near the centre there is a large square stone, crossed by five deep grooves. The circle No. 4 is formed of four blocks of coarse granite, standing nearly on the cardinal points, and about 3 feet high. The figure is not a perfect circle, but rather an ellipsis whose greater axis, directed north and south, is 7 yards long. The circles now described are on that part of the moor which is covered with peat moss, and peat is cut from the banks which are fast approaching the bases of the stones; but this will not affect their stability, as the peat is no more than 15 to 30 inches thick, and the pillared stones have a firm hold in the red till bed below the peat: the peat, indeed, is most probably the growth of a recent time.

The circle No. 5, a little to the west of No. 4, is on higher ground, dry and gravelly, and has remarkable features, which have led to its frequent mention, while others have been overlooked. It is alluded to under the name of "Suithi-Choir-Fhion," or Fingal's Cauldron-Seat; the encircling stones being, according to tradition, the support of the giant's cauldron. It consists of a double circle of granite blocks, boulders from the northern mountains, gathered from the adjoining moor. The outer circle has 14, the inner 8 blocks, larger than those in the outer, and 3 to 4 feet high; they stand 3 to 4 yards apart in the inner row, the diameter of which is 11 yards; the ring between the two circles is from 5 to 7 feet broad, probably because the constructors were in no way particular, or some of the blocks may have been shifted. A block on the south-east side of the outer circle has a ledge perforated by a round hole, which is well worn on the edges; and, according to the usual tradition, is said to have been formed for the purpose of fastening the favourite dog, Bran, belonging to the giant before named. No. 6 is an enclosure by the road side, a short distance W. of No. 5, like an open raised grave; it is formed of five slabs of sandstone placed on edge, 2 to 3 feet above the surface, and so even all round that a large slab may once have fitted on as a lid; the enclosure is 6 feet E. to W., by 4 feet broad; two slabs form the east side, one each of the others. Nos. 7 and 9 are monoliths, but most probably there were circles here also, as there are indications of other stones; the height is about 5 feet 6 inches. South of these is No. 8, having several slabs, nearly the same size as the last, indicating a former circle. No. 10 is a large circle of granite blocks, diameter 21 yards. It is close to Tormore farmhouse, and the most western of the group.

With regard to the nature of the stones of which these circles consist, it is here worthy of note that—while the granite circles consist of such loose blocks of the coarse and fine varieties as are commingled on the moor, and could be readily rolled into position by a number of men with crowbars, none of them

No. 8.

No. 5. Inner diar. 11 yds
Outward do. 15

No. 6. Open Raised Cist.

No. 4.
7 yds diar.

No. 9. Single Stone
4½ Chains West

No. 10. Circle 28
Chains West

No. 1
14 yds diar.

No. 2.
16 yds diar.

No. 3, 13 yds diar.

No. 7. Single Stone.

IMPERIAL CHAINS

1 0 1 2 3 4 5 6 7 8 9 10 11 12 13 14 15 16 17 18 19 20

PLAN OF THE STONE CIRCLES ON MAUCHRIE MOOR, TORMORE, ISLAND OF ARRAN.

Maclure & Macdonald Lith. Glasgow.

being heavier than from 4 to 5 tons—the pillared stones are of Old Red sandstone, which does not occur nearer than the cliffs towards Auchincar, and are of much greater weight, the largest being not less than from 8 to 10 tons, and the others 6 to 8 tons. The intervening country is rough and difficult; yet there does not appear to us any other conclusion possible than that this is the origin whence they have come. All of them have plainly undergone a certain amount of coarse "dressing," but with what tools it is impossible to say.

97. The excavations were begun on the morning of the 24th May, 1861, with a body of nine men, under command of the chief hedger, whose intelligence and zeal were of the greatest use to me. I broke ground in the circle No. 2; and considering that the centre of the circle and the base of a pillar were the spots most likely to receive any object valued or venerated, I found the centre easily, knowing three points on the circumference, and excavated there, and at the base of the most southern of the three pillars, the one on the S.W. side. Nothing was found here; but I may mention, for the guidance of future excavators, that, except in a peaty or gravelly soil, the ground *can be felt* to a considerable depth by means of a strong pole, armed with an iron facing, and thus either much digging be saved or the right direction given to it. The ground was sounded in this way, in the central digging, after 15 inches of peat were cleared off and the till reached, and we became aware, by the peculiar sound, that we were over a flat stone of considerable size. The interest of the inquiry now rapidly increased; and when the earth was cleared off, and a large slab reached, the excitement of our party was wound to a high pitch. There could be no doubt we were upon a place of ancient sepulture, and curiosity was highly excited to discover the contents. The slab being 13 inches thick, was of great weight, and no slight effort, even of our large party, was required to raise it. As it slowly rose, a small, neat chamber was disclosed, but neither skeleton nor bone—nothing but an urn, with a handful of black earth at the bottom; and these

"ashes of the urn" we regarded as all that remained of some great chief, in whose honour and for whose last resting-place these huge monuments had been reared around the central cist. The urn was in excellent preservation, but of rude construction. The cist was 3 feet long, 26 inches deep, and 22 inches broad, formed of four flat sandstone slabs set on edge, the bottom being of the natural hard till soil. In the bottom, lying loosely about in black earth, four flint arrow-heads of rude construction were found. The lid was 2 feet 2 inches from the surface, and fitted nicely on to the sides all round. The length of the cist was nearly N. and S.

The circle No. 3 was next tried at the centre, and a similar cist and urn found at a depth of 4 feet 6 inches from the surface; in the cist and soil over the lid several flint arrow-heads were met with, of rude construction, and the urn was even more rude than the other. The opening was now continued southwards from the centre, and another cist found at a higher level, the surface of the lid being only 20 inches from the surface, the cist itself 2 feet 9 inches; the slab was 4 feet square, and 13 inches thick. No sooner was the northern edge of this ponderous lid slightly raised, so as to give a glimpse of the interior, than the interest of the party was again raised to the highest pitch. A white object, like a blanched human skull, loomed out from the deep obscurity of the cist. We had come at last to a veritable human grave. The skull proved to be tolerably perfect, with most of the teeth entire, the upper jaw partly decayed, the lower only traceable on the floor in outline; at the opposite or north end of the cist, a few long bones lay on the floor, and also two flint arrow-heads. The cist was 3 feet long, 2 feet deep, and 16 inches broad. The dimensions, nearly the same in all the cists, shew that these were not intended for depositories of the body in an extended position. When cremation and the urn were not employed, the body was interred in a contracted position, sometimes on the side with the body and limbs bent; sometimes in a sitting posture, the knees drawn towards the chin, with the arms used in battle lying on either side, ready to be grasped again when the dry bones should be clothed with muscle, and the re-animated body should start into new life!

The double granite circle No. 5 was next tried, but nothing was found save the two sides of a cist. The place had obviously been disturbed before.

The small open enclosure No. 6 was also searched, but nothing met with except a fragment of a flint implement. An unsuccessful search was also made at the monolith on the moor—No. 9 in the plan—and this concluded the labours of our first day; it was bright and warm, and our operations were prolonged late into the evening.

My next examination was made on the 26th September, the force, and its organization by Mr Paterson, being the same as before. His Grace the Duke of Hamilton honoured us with his presence during the greater part of the day, remaining on the ground till nightfall, and aiding us by his advice. The day was one of the most brilliant and genial of the season; the atmosphere had that unusual transparency, ominous of change, which gives magical effects to a landscape. The granite peaks were defined against the deep azure of the north-eastern sky with wondrous distinctness—they seemed close at hand, and lifted up into the clear air far beyond their usual height.

As the circle No. 3 had been found to have a cist to one side of the centre, I was desirous of discovering now whether the circle No. 2, already found, like No. 3, to have a central cist, had also one removed from its centre. The result shewed that it was so; a very perfect cist was found N. of the centre, at 37 inches from the surface; but no object whatever was in it, neither urn, bone, nor arrow-head, nor did it appear to have been ever disturbed; indeed, it is certain it could not have been. The conclusion seems warranted that it had been prepared as the others for a place of sepulture, but never used. It was of about the same size as the others, and like them lay nearly N. and S. To the south of the centre there was no trace of a cist, and it seems therefore improbable that the practice was to form a series round the central one.

Satisfied thus far with the result of the inquiry as regarded the circles of upright stones, I was anxious to determine whether a like arrangement existed in the case of the circles of granite blocks. The cist at the centre appeared to indicate the purpose for which the circumference was reared: and if the central cist exist within the granite circles, the purpose will seem to be the same for both. An imperfect cist had been found in one—two granite circles remained to be examined. The larger of these two, the most eastern of the group, No. 1 on the plan, was tried at the centre, and along a radius, but without success. Nothing was found, although the usual depth was reached. In the other granite circle, No. 4, we were more successful; a beautifully perfect cist was found, 40 inches below the centre, of the usual dimensions, and lying as before, N.N.E. and S.S.W. It contained an urn in fragments, some bone fragments under the urn, three rude arrow-heads, and a small bronze pin, much decayed. This last object added immensely to the interest of the inquiry; while the central cist shews a like purpose in both structures—granite blocks and pillared slabs, alike disposed in reference to a central place of sepulture.

An excavation at the imperfect circle No. 8, high on the moor, gave no result whatever; the position of the centre and course of the circumference were in this case, as in that of No. 9, mere matters of conjecture. This excavation exhausted our day; it was nearly dark when we left the ground, and there was now no time left to examine the last circle, No. 10. I regretted this the less, that it had a tossed appearance as if it had been opened in more than one place, perhaps in the hope of finding treasure concealed in a spot round which there hung a certain amount of sanctity.

Mr Paterson having organised a force at Brodick under the chief forester, next day was given to the supposed site of a circle in Glen Shant, but without any result. The small granite circle and pillared slab at the summit-level of the Lamlash road (Art. 70) were then examined. Here, within the circuit of the four blocks, and at a small depth, I found a cist 26 inches long, 11 wide, and 10½ deep, cut out of the solid sandstone rock, and fitted by a lid. In it there were bone fragments and black earth, in the soil over it some rude flint arrow-heads. At the standing stone we found nothing.

The day turning out very wet and stormy in the afternoon, I was unable to carry out my plan of examining some of the monoliths about Brodick—as those on Mayish farm, and that by the road-side at Invercloy. It would be interesting to determine whether they are true monoliths, or the remains of circles; whether monumental, or commemorative of a battle or the judicial combat. Human remains, or those of weapons, would mark these purposes; the non-existence of any remains would probably merely indicate a boundary or be commemorative of a treaty of amity—that neither party "would pass this pillar for harm" to the other.*

98. All the objects found during the excavations were presented, by the desire of his Grace the late Duke of Hamilton, to the Scottish Antiquarian Society, and may be seen in one group in their Museum at the Royal Society House, on the Mound, Edinburgh. The principal objects are given on two plates. The nature of the human remains was felt to be of great importance in the inquiry—of what race or type of head, of what sex and age, of what bodily proportions, those of a warrior chief, or a tender female, was the individual to whom they belonged? The determination of these questions must have an important bearing on the purpose for which these huge works were erected. Anxious to have the opinion of the highest authority upon these questions, I submitted the entire of the remains for inspection to Dr Allen Thomson, Professor of Anatomy and Physiology in the University of Glasgow, who kindly furnished a full report upon them, which is given at length in the paper already referred to. The remains were

* Genesis xxxi. 51, 52.

also examined by Professor Goodsir, Dr Struthers, and Mr Turner, of Edinburgh. The skull is of the old British type, and, so far, the evidence is in favour of a high antiquity for these works; it seems to be that of a young female, or slender male just arrived at maturity; perhaps "the daughter of the tribe," or the future chief, who it was hoped would one day lead in the field of battle and the chase. The teeth form a nearly complete set; in most of them the enamel, though brittle, was entire, but the bony part crumbled into powder under a very slight pressure; this part was also of a dark brown colour, as if partially charred; the wisdom teeth had either just passed through the gum or were about to do so. Some of the bones were human rib-bones, others shaft-portions of human thigh-bones; there were long bones, portions of deer's horns; and also portions of an animal's under jaw, most probably a dog or seal, but very difficult to determine from the state of decay.

The black earth of the urns was most kindly examined for me by Dr Thomas Anderson, Professor of Chemistry, Glasgow University. It contains many minute bone fragments, small pebbles, sand, and ordinary soil, but no trace of animal matter. The bones, carefully cleaned, have the following composition:—

Phosphates, ...	84·11
Siliceous matter, ...	6·29
Organic matter, ...	3·57
Carbonate of lime, ...	3·41
Water, ...	2·62
	100·00

From this analysis there can be little doubt that the bones have been burned, and that the earthy matter is part of the soil introduced along with them when they were gathered from the spot where the ceremony of incremation was performed. The absence of nitrogen shews that the organic matter is not of animal origin, but that it and the siliceous matter are part of the soil which could not be separated from the bones before analysis.

URN, SKULL, FLINT FLAKES, & BRONZE PIN, FOUND IN STONE CIRCLES, TORMORE.

Found in Cist. N.° 2_ 7" high.

1/3 Nat. Size

Found in same Cist as the Urn

Found in the open Cist. N.° 6.

Found in Cist. N.° 4.

Natural Size.

1/4 Nat. Size.

1/4 Nat. Size.

Lateral View of Skull.

Vertical View of Skull.

Maclure & Macdonald, Glasgow.

OBJECTS FOUND IN THE CISTS.

The urns are of the earliest forms, fashioned by the hand before the potter's wheel was used in these western lands; the irregularity thence arising is very perceptible under the lower band; the form of the mouth also is not perfectly round. The ornamental markings are made by sloping lines, neither straight nor parallel, and appear as if marked in the soft clay with a bit of twig, the streak coming off light towards the lower part; the two rows of alternate dabs appear as if made with the thick end of the twig; the lines separating the bands of ornament are drawn without any attention to accuracy. The height is $7\frac{3}{4}$ inches; diameter, 7 inches at the mouth, $3\frac{1}{8}$ at the bottom; circumference at the lower band, 25 inches; at the upper, 24 inches.

The urn was either sun-dried, like the Mexican adobes, or subjected to but a very slight degree of burning, as Dr Anderson finds that the external surface only is burnt, and the interior appears to have been scarcely heated, for it still contains upwards of 7 per cent. of the water of combination of the clay, which would have been expelled had the urn been strongly heated throughout. The clay has not been brought into a uniform plastic mass, but has the structure of a number of small pellets, with bits of stony fragments, which Dr Anderson suggests may have been used to prevent the urn from cracking during the process of drying and burning. He also considers that the reddish colour of the surface, produced by a peroxidation of the iron, is an additional proof that the urn was subjected to a slight artificial heat.

The flint of which the implements consist is not found in Arran, but it is met with on the opposite coast of Antrim, where it forms beds in the chalk; and thence the rude stone may have been derived, as Cantire and the islands had intercourse with the north of Ireland from the very earliest times. But this is part of a very wide question, and need not be here dwelt upon.

The bronze pin is $2\frac{1}{2}$ inches long, with a polished surface like that of a smooth bone, and resembles the pin of a brooch or bracelet, but slightly thicker. It was chemically tested by Dr Stevenson Macadam, Lecturer on Chemistry, Edinburgh, and by this means, and a comparison with similar objects in the Museum of the Antiquarian Society made by the Secretary, distinctly proved to be of bronze, though much altered in structure.

From the facts which have now been stated, the following inferences seem legitimately deducible:—

1. Whatever may have been the state of civilization among the constructors of these works, a certain sense of fitness or congruity must have existed in their minds; for though there are so many circles, there is no mixture of dissimilar stones—they are either all of sandstone or all of granite.

2. They must have been capable of using mechanical appliances of great power, even on the supposition that the pillared stones are from the carboniferous formation, and were fetched from the nearest point possible—the bed of Mauchrie water.

3. Archæologists divide the pre-historic period in our islands into the Stone and Bronze periods. If their classification be correct, and if it be conceded that there is human progress in every period, then the use of rude flint implements and of implements of bronze ought to be separated by a wide interval, and only flint implements of the most perfect forms, if any, ought to be associated with an article of bronze. But in the present case, flint implements of the rudest forms are associated with an article of bronze; the two have co-existed, have been in use together, and thus the Stone and Bronze periods have interlaced deeply; probably more deeply in an isolated situation such as Arran, than on the adjoining continent of Britain, where improvements in processes of art would spread more rapidly. But we must guard against attaching too much importance to a single case. It is, however, highly desirable that inquiries of this kind should be multiplied in order to test the truth of the theory; it would be greatly invalidated if such association were found to be frequent.

4. All the cists have their greatest length between N. and

N.N.E., and their construction may therefore be inferred to have been anterior to the earlier Christian times in this country, when a superstitious regard began to be cherished for a direction pointing east. The present amount of variation of the compass has been in all cases allowed for; and it does certainly seem strange that the directions should so agree towards a north point—they lie, in fact, roughly N. and S., being all a little E of N., the direction having clearly no sort of connection with the inclination of the surface of the ground, for though this is very various the direction is always the same. Knowing the curious astronomical relations made out by Sir John F. W. Herschel in regard to the pyramids of Egypt, I took the liberty of consulting him on the subject of the direction of the cists within the Arran Stone Circles. He kindly replied that he did not consider that it could have any connection with a past conformation of the heavenly bodies, as affected by the causes of change among them. To refer it to a past amount of variation would of course be absurd. Shall we rather say then that the direction was roughly taken North and South; that it had reference to the mid-day sun, or to a native home of the race to which the constructors belonged, amid the wilds of the north?

5. The information furnished by the human and other bones is not of a very definite kind; the human remains belong to a young person, most probably a female. There are portions of deers' horns, and of an animal's jaw, most probably a dog or seal. The absence of the wanting portions of the human skeleton is not easily accounted for. There was no trace on the floor of the cist of such an amount of matter as the decomposition of the other bones would have left; but if once decomposed, the matter may have been removed, or absorbed by the soil through the floor or spaces between the stones forming the sides, during successive floodings of the cist with water from the soil above. It is highly probable that such removal may have taken place; while the form and position of the skull would preserve it from decomposition for a much

longer time. Had the remains belonged to an aged or powerful male, we should have had much stronger evidence than we now possess for regarding the circumference of huge pillared stones as reared in honour of a great chief or warrior, instead of being a place of mere ordinary sepulture.

6. That these stone circles were erected for the object just stated is further confirmed by the circumstance, that the centre of each circle is marked by a cist; that the circumference has been reared in all cases in reference to this central cist, as well in the case of the pillared stones and granite circles at Tormore, as of the singular little cist, with its four encircling blocks, at the summit-level of the Lamlash road. The limited total area of the circles is no objection to this conclusion, as we know that the sites of many circles, which have existed to a late time, are not now to be found. The circles may have been applied later to other purposes, as of worship, of judicial combat, or of meeting on great public occasions, while intended in the first instance by their constructors as sepulchral monuments, marking off the sacred precincts where lay the ashes or the bones of the dead. The reasoning, it is true, might be reversed, and it might be argued that the circles were reared for religious or judicial purposes, and afterwards adopted for sepulture, as venerated places. But it would, of course, be necessary to produce the evidence for such an original purpose, in order to do away with the force of that here given. This, however, has yet to be done. Antiquarians have too often recourse to hypotheses which are far-fetched, and with which the wonderful, the grand, and the terrible are associated, while simple natural explanations are overlooked or rejected. A Druidical origin has been assigned to these works; but there is no evidence that the Druidical priests, or the rites of that worship, had at any time a footing in Scotland. As regards the south of England, Brittany, and other districts where remains considered to be truly Druidical exist, the soundness of the mode of reasoning just indicated would be tested by a careful examination inside the circles, on the plan described in the

foregoing account. To what extent such an examination may have been made, I am unable to state; within Stonehenge, one trial has, I believe, failed to discover any human remains. Such are, however, found abundantly in the barrows and other earthworks on the adjoining plain.

99. A walk in the late afternoon from Blackwater-foot to Lag, across Leac-a-Bhreac, forms a pleasant close to the day's excursion to King's Cove. Leac-a-Bhreac, or the Brown hill, is a high, round ridge, with steep, seaward front, and long inland extension, lying between Kilpatrick, near Blackwater-foot, and the mouth of Slaodridh water. It is composed of felspar porphyry, being the largest mass of this rock in the island. The base is a gray, or bluish gray, compact felspar, sometimes passing into a jaspideous hornstone, often iron-shot, and the imbedded crystals, glassy felspar; smoky quartz also occurs in round bits, like worn crystals, and the rock affects the columnar form, the columns having flat joints; but the form most frequent is that of rhombic blocks. The surface decomposes into a dark brown layer, from the atmospheric action on the iron of the base. The hills get their name from this circumstance. The rock is well adapted for ornamental purposes, and could easily be procured in any quantity. The porphyry occupies the shore, cutting off the sandstone, and forms considerable cliffs, which exhibit the usual wearing action of the sea when at its old level. The caves are a shelter for the sea spleenwort and other good ferns; and one of the largest is often used as a preaching station in connection with Kilmorie Church. The "grave, sweet melody" of the Scottish psalmody here mingles finely with the mellowed sound of the dashing waves.—On the S.E. side, towards Slaodridh-water, north of the road, the porphyry is seen to rest on sandstone, which, rising gradually from below it, begins to occupy the lower grounds seaward, the porphyry retiring inland, and ranging across the moor to the Crookcrever-burn, where it is associated with clay-stones. Farther up the Slaodridh water, there are again found great beds of a similar porphyry,

as already described in Excursion XV. The great development of these felstones is a remarkable feature of the south-western portion of the island.

Leaving Tormore, or King's hill, and crossing to the Shiskin road, we are in front of the fine-grained granite tract of Glen Iorsa, and accordingly we find that here the great proportion of the boulders are of this variety of granite; while those on the eastern side, and along the southern plateau, are of the coarse variety. This points to a cause locally acting in the direction of the glens: we never find Iorsa granite on the east side of the island, both kinds are mingled on the west side (Art. 22).

If we return to Invercloy on foot, the walk will be varied by passing up Clachan Glen and down upon Glen Cloy. We strike off the main road about a quarter of a mile north-east of Shedok Inn, at the bridge over the Clachan burn, and first examine the curious old effigy in the centre of the small cemetery, popularly called St Molios' Grave, but in reality, we believe, the tombstone of an abbot of Saddell, with his chalice and pastoral staff (Art. 70). The place of worship adjoining is a chapel of ease to Kilmorie Church, near Lag, and has service every third Sabbath; on the two other Sabbaths the hearers attend the Free Church on the west side of the vale, most of them having sittings in that church also—an example of liberal Christian feeling which it would be well to follow. The detrital accumulations at the mouth of Clachan Glen have been already referred to (Art. 90). They consist of the true boulder clay below, with striated stones, and above of the upper drifts, the intermediate shell bed not being present in any of the sections, so far as we could discover; but our examination was not lengthened, and the difficulties are not less formidable than those to be met with in the sections already described (Art. 85). The lower and middle parts of this glen shew the lower portions of the carboniferous series; and high on the south side, apparently by an upthrow of the strata, beds of white limestone, with quartz pebbles, like that of Achab farm and the Fallen Rocks, are brought to the surface. The metamorphic action throughout, but especially towards the head of the glen, is the most extensive and remarkable we have seen in Arran—beds, bands, and dikes of claystone, greenstone, basalt, and syenite, intersect the sandstones and marls, or are intercalated with them, and produce such marked changes that all distinction between sedimentary and igneous products is obliterated. The sandstones pass across the watershed into Lamlash Glen on one side, and Glen Cloy on the other.—From the high plateau above Ploverfield, we may descend either by Windmill Hill or the Sheeans. To walk from Invercloy over the "String" by Mauchrie waterfoot and Tormore, and return by this route, is no mean performance for one day. But the platform ought to be gained before sunset, as twilight falls rapidly in Glen Cloy, and the ground between the Sheeans and Invercloy is broken and difficult.

FOSSILS OF THE ARRAN ROCKS

100. The following is a list of all the fossils yet discovered in the various sedimentary formations of Arran, corrected up to the present time. Most of the fossils in the list were collected and named by Mr John Young, Hunterian Museum, Glasgow University, and Mr James Thomson, an active member of the Glasgow Geological Society. Mr Nelson Mitchell and Mr Thomas Chapman have kindly supplied several good species. The Corriegills' orthoceras could not be specifically determined, on account of the change produced by a trap dike on the sandstone in which it was imbedded. The list of Arctic shells was kindly supplied by the Rev. Robt. Boog Watson of Edinburgh, their first discoverer, who also, in the most prompt and kind way, placed at my disposal all the information in his power in regard to the localities whence the shells were obtained. The Rev. H. W. Crosskey selected for me the foraminifera found in the shell bed, and forwarded them to Prof. Rupert Jones, to whose kindness I am indebted for the names of the species. Several others since found in the shell clay I have not been able to get named in time.

THE SLATES.

No trace of any fossils.*

THE OLD RED SANDSTONE.

Markings on the South Sannox beds look like fucoid or annelid impressions, but they are not decidedly organic.

* Yet, if the views of Sir R. Murchison in regard to the slates of the north-west of Scotland be correct, these slates must be regarded as of Silurian age. Metamorphic action may have obliterated all trace of fossils.

THE CARBONIFEROUS SERIES.

Plantæ.	
Stigmaria ficoides (sandstone), . .	Corrie, Salt-pans.
Calamites cannæformis, " . .	———
Sphenopteris (?), ...	Salt-pans.
Lepidodendron obovatum, . . .	———
Zoophyta.	
Cyathophyllum Wrightii, . . .	Corrie, Salt-pans.
——— expansum, . . .	Corrie, Salt-pans.
Cyathopsis eruca, ...	Corrie.
Lithodendron junceum, . . .	Corrie, Salt-pans.
Chætetes tumidus, ...	Corrie, Salt-pans.
Echinodermata.	
Archæocidaris Urii, ...	Corrie, Salt-pans.
Poteriocrinus crassus, ...	Corrie, Salt-pans.
Hydreionocrinus globularis, . . .	Salt-pans.
Crinoid stems, various, . . .	———, Corrie.
Crustacea (Trilobites).	
Griffithides mesotuberculatus, . .	Salt-pans.
Polyzoa.	
Ceriopora interporosa, ...	Salt-pans.
Fenestella plebeia (?), . . .	———
——— oculata, ...	———
——— Morissii, ...	———
Diastopora megastoma, . . .	———
Polypora papillata, ...	———
Brachiopoda.	
Terebratula hastata, ...	Salt-pans.
Spirifera bisulcata, ...	Corrie, Salt-pans.
——— lineata, ...	———

Spirifera Urii, ...	Corrie, Salt-pans.
——— duplicicostata, . . .	———
Spiriferina octoplicata, . . .	Corrie, Salt-pans.
——— laminosa, ...	———
Athyris ambigua, ...	Salt-pans.
——— Royssii, ...	———
Retzia radialis, ...	———
Strophemena rhomboidalis, var. analoga,	———
Rhynchonella pleurodon, . . .	Corrie, Salt-pans.
Streptorhynchus crenistria, . .	Salt-pans.
Orthis resupinata, ...	———
Chonetes Hardrensis, ...	Corrie, Salt-pans.
Productus giganteus, . .	Corrie, Salt-pans, Glen Shirag, Mauchrie water.
——— semireticulatus, . . .	Corrie, Salt-pans.
———, var. Martini,	———
——— longispinus, . . .	———
——— latissimus, . . .	———
——— punctatus, ...	———
——— costatus, ...	———
——— scabriculus, . . .	Salt-pans.
——— aculeatus, ...	———
——— Youngianus (?), . . .	———

Lamellibranchiata.

Monomyaria Dimyaria.

Aviculopecten fimbriatus, . . .	Corrie, Salt-pans.
——— several undetermined.	
Myalina Verneuilii, ...	Corrie, Salt-pans (?).
Venus (?), two sp. undetermined, .	Salt-pans.
Edmondia sulcata, ...	———
Nucula gibbosa, ...	———
Leda attenuata, ...	———
Conocardium alaeforme, . . .	———
Cypricardia rhombea, ...	———

Gasteropoda.

Pleurotomaria Yvanii, . . .	Salt-pans.
Macrocheilus acutus, ...	Corrie.
———— Michotianus, . . .	——
———— fusiformis, . . .	Salt-pans.
Naticopsis elliptica, ...	Corrie.
——— variata, ...	——
Murchisonia striatula, . . .	Salt-pans.
Trochus biserratus, ...	——
Loxonema scalaroidea (?), . . .	——
Bellerophon Urii, ...	——
——— striatus, ...	Corrie.

Cephalopoda.

Orthoceras unguis, ...	Corrie.
——— attenuatum, . . .	Salt-pans.
——— (?), ...	Corriegills, in sandstone.
Nautilus ingens, ...	Salt-pans.
——— tuberculatus, . . .	——

Pisces.

Megalichthys Hibberti, . . .	Corrie.
Cochliodus magnus, ...	Corrie, Salt-pans.

Newer Pleiocene, or Arctic Shell Beds.

Balanus crenatus.
Panopæa Norvegica.
Tellina Baltica.
Cyprina Islandica.
Astarte elliptica.
——— Arctica.
——— compressa.
——— striata.
Cryptodon Sarsii.
Modiola modiolus.
Leda pygmæa.
—— pernula.
Pecten Islandicus.
——— opercularis.
Littorina littorea.
Turritella communis.
Natica (?).

Foraminifera and Entomostraca, in the Glacial Clay.

Rotalia Beccarii.

Polystomella striato-punctata.

Cythere, a new species, closely allied to C. Lamarckiana, an Atlantic living form undescribed.

THE FLORA OF ARRAN

101. In the preceding pages it has been shown how fine a field Arran offers to the geologist, as exhibiting many diversified phenomena in a limited area. To the botanist it is scarcely less interesting—such is the luxuriance and variety of its vegetation, and such the rarity of some of the plants contained in its flora. It is, indeed, true that scarcely any of these are botanical treasures of the first order; still there are several of unfrequent occurrence in the west of Scotland, and many quite new and highly interesting to the naturalist accustomed to the flora of England or the Lothians.

This richness of Arran as a botanical field is owing to two causes—its geographical position and the variety it affords of situation and soil. Lying near the shores of the Scottish Lowlands, and at the same time forming one of the Hebridean chain of islands, it partakes of the flora of each region—the common plants of its fields, woods, marshes, and road-sides belonging chiefly to the former—the maritime species to the latter. These advantages of position it no doubt does in some measure share with the coasts of the Clyde estuary generally, and especially with Bute; in Arran, however, they meet most completely; and accordingly we find that no district of equal extent in the west of Scotland can rival it in the number of species. It is to this circumstance—its situation at the junction of two dissimilar botanical provinces—that the peculiar richness and variety of the flora of the island are mainly to be ascribed.

Secondly, The geological structure of Arran impresses a marked character on its physical geography, and gives rise to the greatest possible variety of station and soil. We have lofty and precipitous mountains, wide-spread moors, small alluvial plains, hot and sheltered glens, damp woods, and sandy sea-shores. We have every kind and degree of exposure, from the wind-swept top of Goatfell to the warm hollows of Glen Cloy, and this within the compass of a few miles. We have modern fir plantations and natural birch woods; cultivated fields and hedge-rows; wide stretches of peat-bog; rocky promontories and caves; open strands and sand hills. Of soils, too, in the stricter sense of the word, there is a notable variety. The general division of earths into sandy, loamy, clayey, and so forth, is loose, and for scientific purposes inaccurate; distinctions far more natural, as well as valuable, are furnished by the nature of the rocks whose decomposed materials form the soil. In Arran the variety of the geological formations produces a corresponding diversity in the composition of the earths; and though we are no doubt still greatly in the dark as to the influence of soil on the habitat of plants, there can yet be little doubt that the diversity of soils furnished by the granites, traps, porphyries, slates, sandstones, shales, and limestones, must exercise a powerful influence on the flora. Whatever may be the importance of these conditions generally, they certainly exist here in a remarkable degree.

To these advantages of geographical position and physical character is to be added the scarcely less important one of climate. Lying off the west coast of Scotland, and thus almost in the Atlantic, Arran enjoys a copious rain-fall, while its insular position preserves it from those extremes of heat and cold which are so injurious to vegetation. Hence it is that the climate, while in many parts quite cold enough for the ordinary plants of Britain, and while the mountains rise high enough to be a fit habitation for several alpine species, is yet sufficiently mild and equable for the growth of some usually found in more southerly regions. These conditions exist also in the extreme point of Cantire; and accordingly we find Campbelton enjoying the mildest winter in Scotland, and many tender species flourishing there in the open air.

The effect of climate on the flora of Arran is twofold—it increases the number of species, and it imparts a general vigour and luxuriance to the vegetation, which makes it interesting to the lover of the landscape picturesque, no less than to the botanist. Every one must have observed how much of the peculiar charm of Arran scenery is owing to this circumstance. The bright green of the fields and pastures, the woods that fringe the shore and cling to the slopes of the lower hills, add grace and soft beauty to landscapes that would otherwise be severe and gloomy in their grandeur; and while they pleasingly relieve the monotonous gray of the granite mountains, serve to heighten by contrast the effect of the bare crags and jagged peaks that rise behind them. In spring or early summer, when the grass sprouts fresh on the hill-sides, and the varied foliage of the trees still preserves the delicate green of youth, or in July, when the lower ridges are purple with the rich heather-bloom, this variety of colour is finest—and this is the best time to see Arran scenery in its perfection. In September the grass is parched, the heather begins to wither, and a brown or gray tint prevails over the whole landscape, little relieved even by the red and yellow hues of the autumnal trees. While, again, if we turn our attention from the general outlines of the scene, and look more closely at its details, we shall find no less cause to admire the effects of an atmosphere moist and warm. To it is owing not only the abundant growth of natural wood which clothes the glens and sea-cliffs, but also the size and splendour of many of our finest shrubs and wild flowers, such as the laburnum, the hawthorn, the foxglove, the honeysuckle, and perhaps most strikingly of all, the luxuriant profusion of the whole fern tribe, from the stately *Osmunda* to the lowly *Hymenophyllum*.

102. On proceeding to analyze the flora of a given district, that is, to examine the internal relations of its constituent

parts, noting what species, or genera, or families, are abundant, what rare, and what character is thus imparted to the vegetation, we may regard it under two aspects, as illustrating the peculiar features of the tract itself, or as indicating the relations of that tract to the surrounding regions—what may be called its general botanico-geographical position. Glancing at the flora of Arran from this latter point of view, we shall observe some interesting facts. Writers on the botanical geography of Britain class the plants of our country according to several "types of distribution," to denote their geographical range and affinities. Thus many species, occurring chiefly in the east and south of England, are assigned to the Germanic type; others, most of them denizens of the Scottish Highlands, are referred to the Scandinavian; while others again, characterising Ireland and the westerly coasts of Britain, are grouped under the Atlantic type. Some few there are, found only in Cornwall and the west of Ireland, whose affinities are with the Spanish Peninsula, and especially with the mountains of the Asturias, and for these there is proposed a Lusitanian type. Somewhat similar to the position of Britain, if we may be allowed to compare great things with small, is the position of Arran. Most of its common species—the plants of the field, the road-side, the marsh—it has in common with the western Lowlands of Scotland, which stand to it in the same relation that the continent does to England. These Lowland plants form the bulk of its flora. They include almost all those of common occurrence, as well as several of the rarer sort—such as *Ranunculus Lingua, Helianthemum vulgare, Epipactis ensifolia, Samolus Valerandi,* the *Botrychium* and *Ophioglossum, Asplenium ruta-muraria,* etc. Under this class are embraced nearly all the species that frequent the cultivated land, the marshes and streams, the woods and pastures; together with several maritime ones—as *Silene maritima, Oenanthe Lachenalii, Calystegia Soldanella.*

A second "type" discernible in Arran is that which we might call the Highland. To this group belong the alpine plants of the highest granite mountains—*Salix herbacea, Thalictrum alpinum, Alchemilla alpina, Cryptogramma crispa,* and others; several also occurring in elevated situations in various parts of the island—as *Rhodiola rosea, Oxyria reniformis, Hymenophyllum Wilsoni;* besides a few found in mountainous regions, though at no great altitude—such as *Corydalis claviculata* and *Polypodium Dryopteris.*

We have, thirdly, a considerable class occupying, as it were, in Arran, the place which the Atlantic type holds in Britain, including the plants peculiar to the west coast, and especially frequent in the Hebridean chain of islands. Such are many of the maritime species—*Mertensia maritima, Brassica Monensis, Sedum Anglicum, Raphanus maritimus,* as well as *Pinguicula Lusitanica, Gymnadenia conopsea, Drosera Anglica,* and *Listera cordata*—plants found in various localities through the interior. This class includes many of the most interesting and characteristic plants of the island—not a few of them such as will be entirely new to the English botanist.

But the most curious feature in the botanical geography of Arran is the occurrence in its southern extremity of several species scarcely elsewhere to be found in Scotland; belonging, in fact, to the flora of central England, and here apparently quite projected, so to speak, from their ordinary range. Of these the most remarkable are *Lathyrus sylvestris, Verbascum thapsus, Inula Helenium, Althœa officinalis,* and *Carlina vulgaris.* They all occur within the circuit of a mile, on the warm southern face of the cliffs and steep alluvial banks that front the sea at the extreme south of the island, near Benan-head. No one who examines the locality can think it possible that they should have escaped from cultivation; and it is scarcely less improbable that they should have been planted there by the hand of man. The *Lathyrus, Verbascum,* and *Carlina* are still abundant; the *Inula,* however, seems to have been extirpated, if indeed the report of its existence was correct. It is certainly not to be found now, yet it is difficult to see how the mistake could have arisen, as there is no plant in the neighbourhood which would be readily mistaken for it. The *Althæa* has been found within the last few years, but is now either extinct or very scarce. It is to be hoped that botanical collectors who may visit the spot will spare the beautiful *Lathyrus*, and refrain from extirpating it in what is probably its only Scotch locality.

103. Having remarked thus on the external relations of Arran to the botanical geography of the surrounding lands, we might pass on to speak of its internal aspects, and notice the several floras of the mountain, the glen, and the shore, the moor, the wood, and the marsh; in other words, to distribute the plants according to their respective botanical *stations*. By such an examination, however, no new facts or principles of peculiar interest would be elicited. A little experience will enable the student to frame such a classification for himself.

Most of the rare plants of Arran are to be obtained on or near to the coast, some decking the bright sands, as the *Brassica Monensis*, the purple *Mertensia*, or oyster plant, as it is called from the flavour of its leaves, and the lovely *Calystegia Soldanella*, with its creeping stems and flowers of delicate pink; some dwelling in the salt marsh and wet grounds that lie between the old sea-cliff and the present tide-mark, such as the *Oenanthe Lachenalii*, *Triglochin maritimum* and *palustre*, the pretty little *Glaux*, the blue *Aster*, and several others. Here, too, though not properly maritime plants, we often find the handsome *Parnassia*, and delicate *Anagallis tenella*, *Samolus Valerandi*, *Orchis latifolia*, and, in the drier spots, *Geranium pratense* and *Erythræa linearifolia*. The bare rocky crags and promontories, which here and there diversify the generally accessible coast of the island, are gay with the brilliant white, yellow, and pink flowers of *Sedum Anglicum* and *acre*, *Silene maritima* and *Spergularia marina;* the succulent *Cotyledon* fixing its roots in the rock clefts, and the glossy green of the sea spleenwort, *Asplenium marinum*. The curious sea-cliff which lines the coast of the island in almost every part, marking the level at which the sea stood in some former age, is in most places thickly covered with a natural growth of oak, ash, birch, hazel, and other trees, and is kept moist by the numerous streams that trickle down its face, or precipitate themselves in cascades from its edge. The shade and humidity thus produced render its vegetation luxuriant and varied; and we find many interesting species growing on or near this line of irregular cliff, some hanging from its wooded sides, some springing rank in the wet caves that pierce it, and some inhabiting the stony and marshy ground at its base.

Here, among others, occur *Veronica montana*, *Hypericum Androsaemum*, *Geranium sanguineum*, *Sanicula Europaea*, *Eupatorium cannabinum*, *Lycopus Europaeus*, *Listera ovata*, *Habenaria chlorantha*. Here, too, most of the Arran ferns may be found—*Polypodium Dryopteris* and *Phegopteris*, *Cystopteris fragilis*, *Lastrea recurva*, *Aspidium lobatum* and *aculeatum*, *Asplenium marinum*, *Hymenophyllum Tunbridgense* and *Wilsoni*, the rare *Trichomanes radicans*, and the magnificent *Osmunda regalis*. In naming these plants, we do not mean that they are all to be met with in any one spot, but that they all occur in some part or other of the sea-cliff, while many abound through the whole of its long extent. Those species which have just been mentioned, with the addition of one or two rarities, such as *Epipactis ensifolia* and *Thalictrum flavum*, form in the main the sylvestral flora of the island, which it is therefore needless to speak of more particularly. Similarly, there is little to distinguish the vegetation of the lower glens from that which has been described as characteristic of the sea-cliff and the woods, not, at least, till the point is reached where the larger trees grow scarce, finally giving place to thickets of birch or hazel, or to the open expanse of pasture and moor. Here the aspect of the scene is changed, and plants quite different attract the attention of the botanist. The greensward is gay with the purple *Gymnadenia conopsea*, as beautiful as it is fragrant, the blue *Jasione montana*, *Pimpinella saxifraga*, *Gentiana campestris*, *Erythraea centaurium*, *Habenaria viridis* and *albida*, *Orchis maculata*, and many other handsome plants. Nestling among the heather we may find *Circaea alpina, Listera cordata,* with its slender stem and minute yellowish flowers, the taller *Galium boreale,* and the tender green of the oak fern, *Polypodium Dryopteris. Rubus saxatilis* here and there trails its long stems over the stony ground, while the viscid leaves of the sundew, *Drosera rotundifolia* and *Anglica,* the tiny cream-coloured flowers of *Pinguicula Lusitanica,* and straggling yellowish-green stems of the little *Lycopodium selaginoides,* mix with the moss that grows thick round the margin of the springs and rivulets. Supposing the glen, whose botany we have been describing, to be in the southern division of the island, we shall, on ascending still higher, find ourselves, after a stiff climb, on a wide expanse of undulating moorland, covered by a thick deposit of peat bog, interrupted here and there by a rocky hill-top, or the deep cut channel of some mountain burn. These moors, varying in height from 900 to 1,400 feet, have little to interest the botanist. He may travel over them for a whole day without meeting more than two or three species among the coarse grass and heather, mixed with rushes and cotton grass, which clothe the surface of the peat moss. Generally it may be said, that the flora of the higher grounds in Arran is inferior to that of the low country. The granite mountains in the northern part of the island rise quite into the alpine region, and are covered with snow for seven or eight months in the year. Yet, when compared with the mountain tracts of the central Highlands, they will be found to possess few alpine plants. This fact seems to be in some measure owing to that predominance of bare rock over grass and heather, which gives them, at some distance, the appearance of an unbroken mass of gray granite. Besides, the stony character of the mountain slopes, the thinness and ungeniality of the soil furnished by the decomposed granite, the absence in the higher regions of springs and streamlets, and perhaps the very steepness of the loftier summits, may all contribute to render the alpine flora of Arran comparatively scanty and uninteresting. But as this poverty is hardly more marked in Arran than on the other Clyde mountains and through the south-western Highlands generally, it ought perhaps to be chiefly ascribed to the distance of these regions from that centre of distribution whence we suppose the alpine species of Scotland to have spread themselves. The commonest plants on the high mountains are *Saxifraga stellaris* and *Alchemilla alpina;* the former is scarce on Goatfell, occurring more abundantly on the heights round the head of Glen Sannox; the latter is very frequent on all the higher peaks, and covers, with the graceful drapery of its silky leaves, the ledges of many a granite precipice. *Salix herbacea*—the dwarf willow, whose woody stem scarcely rises from the ground—is found on most of the principal summits; *Oxyria reniformis* and *Rhodiola rosea* grow abundantly in the rock clefts; *Circaea alpina* and *Saxifraga hypnoides* occur occasionally near the summits of the southern hills; the pretty little *Thalictrum alpinum* may be found in many places, as on Ben-Varen, on Goatfell, at the head of Glen Cloy, and on the summit of the pass leading from Glen Rosa into Glen Sannox; *Cryptogramma crispa*, the parsley fern, has been noticed in several spots, rooting deep among the loose blocks of stone that strew the mountain-side, while among the other crytogamous plants, *Polypodium Dryopteris* and *Phegopteris, Cystopteris fragilis, Hymenophyllum Wilsoni,* and several of the alpine *Lycopodia,* may be enumerated as denizens of the glens and mountains.

Regarding the water plants of the island, those which we find in its marshes and streams, there is but little to be said. They are few in number, and not in any way remarkable. The list is nearly exhausted by the names of *Hypericum elodes, Ranunculus Lingua, Drosera Anglica, Littorella lacustris, Alisma plantago* and *ranunculoides,* and several species of *Potamogeton* and *Carex.* Of true lake plants there are very few; and this fact suggests the remark, how deficient Arran is in this element of the picturesque. Lakes there are several, but, with scarcely an exception, they are placed in the high bleak moors, far above the limit to which trees ascend, and generally away from the higher mountains, so that they add little either of beauty or of sublimity to the scenery of the island. Loch Tanna, by far the largest, lies in a high and gloomy plateau strewed with blocks of decomposing granite, among which the stunted heather barely supports its existence—itself as black and uninteresting a sheet of water as any pool in the fens of Lincolnshire; the aspect of the whole scene is one of utter wildness and desolation, without grandeur. The absence, however, of this element of the beautiful in Arran scenery is scarcely remarked, since it is more than compensated for by the character which the sea imparts to every landscape.

104. Having thus briefly indicated the plants which the botanist will meet with in each region of the island, it is scarcely necessary to prescribe for him any special excursion or walks. These are best left to his own taste and convenience. It may, however, be not amiss, in a few concluding words, to direct him to the districts where his rambles will be attended with most pleasure and success. Probably no part of the island will offer to him so many interesting species as the vicinity of Brodick, especially if he direct his walks to the Corriegills shore, proceed northwards to Corrie, or explore the tangled thickets and dripping rocks at the head of Glen Cloy. Around Loch Ranza, too, several excellent plants may be obtained; while, even if the tourist be not botanically inclined, he will find in the exceeding beauty of the coast an ample reward for his walk along the lovely shore between Glen Sannox and Newton Point. The western coast offers many striking scenes, and everywhere commands noble views of the broad sound of Kilbrannan, with the hills and glens of Cantire beyond; but its botany presents little that is new to one who has already examined the eastern part of the island; while the interior of the country is occupied by undulating granite mountains, seldom, except at Glen Catacol, assuming forms of sublimity or beauty, and clothed with no vegetation beyond the grass and heather that grow among the slowly-decomposing blocks of gray granite with which the ground is strewed for

miles. Such is the aspect of the country—bleak, wild, unvaried—from Loch Ranza to Loch Iorsa and Dugarry, where the slate and old red sandstone formations succeed the harsher granite, and subside with gentle declivities into the alluvial plain through which the Mauchrie water finds its way to the sea. Despite what has been said of the botanical attractions of Brodick, there is no district in Arran that will better repay the trouble of a visit than the south coast, from Slaodridh to Whiting Bay. Without any of the alpine grandeur of the north, it has many striking beauties of its own; smiling little bays, steep green banks, and bold cliffs of basaltic rock, trap-porphyry, or claystone, jutting far out among the waves, or running in tall colonnades along the shore; seaward there is the wide expanse of glorious blue, with the magnificent pinnacle of Ailsa full in front; beyond all, closing the distant horizon, the gleaming cliffs of Ayrshire, and the far-off coast of Ireland. It is a delightful shore to wander along slowly, searching and prying for rarities in the salt-marsh by the water's brink, or up some leafy gorge through which the streamlet from the hills forces its seaward way, forgetting the world without and all its cares, delivered from the dominion of dusty roads, and the necessity of getting home in time for dinner. Let the naturalist take up his quarters in the neat little inn that lies nestling in its snug little hollow at Lag; there he may pass happy days in exploring that solitary shore, and at night pull out into the deep, and taste the unwonted pleasures of sea-fishing.

The mountains at the head of Glen Sannox will be found richer in alpine plants than Goatfell, though scarcely equalling it in height. Yet even their flora must appear scanty and uninteresting to one who has botanised over the ranges of the central and eastern Highlands, Ben-Lawers, Braemar, or Clova. But at this he will have no cause to repine when he finds himself led into some of the most magnificent mountain scenery in Britain. It has been said that no scenery in Scotland but that of Coruisk in Skye, and that of Glencoe, can rival the grandeur of Glen Sannox; certainly neither the Perthshire Highlands, nor the English lake country, can shew anything so wildly sublime as the precipices and aiguilles of rifted granite which tower around the heads of these Arran glens.

LIST OF PLANTS.

105. The following list comprises only the rarer plants of Arran, including under this title both those which, though frequent in the west of Scotland, are seldom found in other parts of the country, and those which, while abundant in other parts of Britain, may appear somewhat uncommon to one who knows only the botany of the west coast. It often becomes difficult to draw the line, and to say what plants should be omitted and what inserted. We cannot hope to have decided always judiciously; when we have erred it has no doubt been in admitting plants which the practised botanist will think too common to be placed in any but an exhaustive list.

All the species of the fern tribe known to exist in the island have been inserted, and their localities given in full detail; partly because they bear a larger proportion than usual to the flora of Arran, and partly for the sake of those who may take a special interest in them, or wish to procure specimens for cultivation.

The genera are arranged according to the natural orders, and the names those which appear in Hooker and Arnott's *British Flora*.

Our best thanks are due to Professor J. Hutton Balfour of Edinburgh, for his kindness in placing his lists at our disposal. On the authority of these many species have been inserted. A few not observed either by Dr Balfour or ourselves are inserted on the authority of the late Dr Landsborough. To these the letter L is affixed. Plants which we suspect to be not truly wild are indicated by an asterisk. It may be added, that as such a catalogue cannot profess to be complete, any hints for its correction or extension will be gratefully received.

Thalictrum alpinum, .	Goatfell, head of Glen Sannox, Ben-Varen, head of Glen Cloy.
——— flavum, .	Whiting Bay (L).
——— minus, .	Whiting Bay.
Ranunculus hederaceus, .	In many places.
——— Lingua, .	Lamlash.
——— acris (var. pumilus)	Near Lamlash.
——— sceleratus, .	
Trollius Europæus, .	
Corydalis claviculata, .	House roofs and woody places.
Cardamine hirsuta, .	Near Lamlash.
Cochlearia officinalis (several varieties), . .	On the shores.
Cakile maritima, . .	Frequent on sandy shores.
Lepidium Smithii, . .	Brodick.
——— campestre, .	Loch Ranza (?).
Sisymbrium Sophia, .	Sandy sea shores.
Brassica Monensis, . .	Sands at Brodick, Sannox, Black-water-foot, etc.
——— oleracea, . .	Lag.
——— campestris, .	Mauchrie.
Crambe maritima, . .	Imochair.
Raphanus maritimus, .	Southend.
Helianthemum vulgare, .	Kildonan.
Viola tricolor (var. arvensis),	
——— palustris, .	Marshy places.
Drosera Anglica, . .	Frequent.
——— rotundifolia, .	Do.
Parnassia palustris, .	Do.
Silene maritima, . .	Shores—frequent.
Lychnis diurna, . .	Common.
——— vespertina, .	Do.
Agrostemma Githago, .	Cornfields—common.
Sagina maritima, . .	On the shores.
——— nodosa, . .	Lamlash.
Honckenya peploides, .	Sandy shores.
Stellaria media, . .	Frequent.
Spergularia marina, .	On the coast.
Radiola millegrana, .	Goatfell, Loch Ranza, Springbank.
Althæa officinalis, . .	Struey Rocks (?).
Malva sylvestris, . .	
Hypericum perforatum, .	Frequent.

Hypericum dubium, .	Invercloy, Whiting Bay, etc., abundant.
——— Androsæmum,	Frequent in thickets.
——— quadrangulum,	Wet places.
——— humifusum, .	Frequent.
——— pulchrum, .	Common.
——— hirsutum, .	Lamlash.
——— elodes, .	King's Cove and Loch Ranza.
Geranium pratense, .	Brodick, Benan-head, Holy Isle, etc.
——— sanguineum, .	Struey, Thundergay, Dippen.
——— dissectum, .	Frequent.
——— pusillum, .	Do.
Trifolium filiforme, .	
——— medium, . .	
——— procumbens, .	
Anthyllis vulneraria, .	Frequent.
Lotus major, . . .	Common.
Orobus tuberosus, . .	In the woods.
Vicia sylvatica, . .	King's Cove, Kildonan, etc.
—— hirsuta, . .	
—— Cracca, . .	Frequent.
—— sepium, . .	Do. (?).
Lathyrus sylvestris, .	Struey Rocks.
Geum intermedium, .	Frequent.
—— urbanum, . .	Common.
—— rivale, . .	Frequent.
Comarum palustre, .	Frequent.
Alchemilla vulgaris, .	Common.
——— alpina, . .	On the mountains.
——— do. (var. conjuncta)	Glen Sannox.
——— arvensis, .	Frequent.
Agrimonia Eupatoria, .	On the south coast.
Prunus communis, . .	
Rosa spinosissima, . .	Frequent.
—— canina, . .	Common.
—— involuta, . .	Lamlash.
—— tomentosa, . .	Dippen.
—— villosa, . .	
Pyrus aucuparia, . .	Frequent.
—— pinnatifida, . .	Glen Eais-na-vearraid, Caistael Abhael
Rubus carpinifolius, .	Lamlash.
——— corylifolius, . .	Do.
——— Idaeus, . .	Woods and mountains—frequent.
——— saxatilis, . .	Head of Glen Cloy.
——— suberectus, . .	Holy Island and Lamlash.
——— glandulosus, .	
——— rhamnifolius, .	
Circæa Lutetiana, . .	Frequent.

Circæa alpina, . .	Glen Cloy, Bein Leister Glen.
Peplis Portula, . .	
Lythrum salicaria, . .	Frequent.
Montia fontana, . .	
Scleranthus annuus, .	In many places.
Cotyledon umbilicus, .	Frequent.
Sedum Rhodiola, .	Mountains—frequent.
—— *Telephium, .	
—— acre, . . .	Frequent on the shores.
—— Anglicum, . .	Common.
Saxifraga stellaris, .	Mountains—frequent.
—— hypnoides, .	Glen Cloy, etc.
Hydrocotyle vulgaris, .	Common.
Æthusa Cynapium, .	Waste ground.
Sanicula Europæa, .	Thickets.
Eryngium maritimum, .	Sandy shores.
Bunium flexuosum, .	Frequent.
Helosciadium nodiflorum,	Shore at Leac-a-Bhreac.
—— do. (var. repens),	Lamlash.
Carum *Carui, . .	
Pimpinella Saxifraga, .	Frequent.
Oenanthe Lachenalii, .	Corriegills, etc.
Conium maculatum, .	Dippen, Lag.
Symrnium olusatrum, .	Dippen (L).
Scandix Pecten, . .	Frequent.
Myrrhis odorata, . .	Dippen.
Ligusticum Scoticum, .	Southend, Kildonan, etc.
Apium graveolens, .	Shore at Lag, and Loch Ranza.
Galium boreale, . .	Glen Laodh, North Sannox, and other places.
Asperula odorata, . .	Woods.
Valeriana officinalis, .	Frequent.
Fedia dentata, . .	Loch Ranza.
Cnicus pratensis, . .	
Carlina vulgaris, . .	Struey Rocks.
Carduus palustris, . .	
Bidens cernua, . .	Brodick.
—— tripartita, . .	Lamlash.
Inula Helenium, . .	Struey Rocks (?).
Eupatorium cannabinum,	In many places.
Antennaria dioica, .	Pastures and moors.
Cichorium Intybus, .	Whiting Bay.
Gnaphalium uliginosum,	Loch Ranza.
—— sylvaticum,	Loch Ranza.
Aster Tripolium, . .	Salt marshes.
Hieracium murorum, .	Near Lamlash.
—— sylvaticum (var. vulgatum),	Glen Eais-na-vearraid.

Filago Germanica, .	Brodick, Glen Eaisdale.
——— minima, . .	Glen Eaisdale.
Solidago virgaurea, .	Glens—frequent.
Senecio sylvaticus, .	Lamlash.
Pulicaria dysenterica, .	Struey Rocks.
Matricaria maritima, .	Sea coast.
Jasione montana, . .	Common.
Lobelia Dortmanna, .	Small loch near Loch Ranza; Loch Iorsa.
Vaccinium Myrtillus, .	Frequent.
——— Vitis-Idæa, .	On the higher mountains.
Arctostaphylos Uva-ursi,	Holy Isle, Glen Eais-na-vearraid.
Pyrola minor, . .	Holy Island (L).
Ilex Aquifolium, . .	On the cliffs.
Fraxinus excelsior, . .	Rocks at Glen Catacol.
Erythræa Centaurium, .	Frequent.
——— linearifolia, .	Shore at Corriegills, etc.
Gentiana campestris, .	Moors and pastures.
Menyanthes trifoliata, .	Springbank, etc.
Convolvulus arvensis, .	Corrie.
Calystegia sepium, .	Hedges—frequent.
——— Soldanella, .	Blackwater-foot.
Cuscuta *epilinum, .	Lamlash.
Lithospermum officinale,	Loch Ranza.
——— arvense, .	
Mertensia maritima, .	On the sands, Brodick, Sannox, etc.
Solanum Dulcamara, .	Brodick, Holy Island, etc.
Veronica montana, .	Blue Rock, and woods in other places.
Melampyrum pratense, .	Abundant.
Scrophularia aquatica, .	Near Sannox.
Digitalis purpurea, . .	Frequent.
Linaria vulgaris, . .	
Verbascum Thapsus, .	Cliffs at Struey and Dippen.
Lycopus Europæus, .	Sannox.
Mentha sylvestris (var. velutina)	
——— sativa (var. rubra),	Lamlash and Whiting Bay.
——— viridis, . .	Near Corrie.
——— *rotundifolia, .	Brodick.
Galeopsis versicolor, .	Frequent.
Calamintha Clinopodium (?),	
Lamium amplexicaule, .	Loch Ranza.
——— incisum (?), .	Do.
——— intermedium, .	Lamlash, Kildonan, Mauchrie, etc.
Stachys ambigua, . .	Slaodridh.
Scutellaria galericulata, .	Brodick, Struey, etc.
Origanum vulgare (?), .	
Pinguicula vulgaris, .	Abundant.

Pinguicula Lusitanica, .	In the bogs, but now less frequent than it was formerly.
Utricularia vulgaris, .	Loch-an-Deavie, Loch Iorsa.
———— minor, .	Loch on the shore, near Mauchrie.
Glaux maritima, . .	Salt marshes.
Anagallis arvensis, . .	In cultivated ground.
——— tenella, . .	Abundant in the bogs.
Samolus Valerandi, .	Marshes—frequent.
Armeria maritima, . .	Sea shore—abundant.
Littorella lacustris, .	Loch Ranza.
Plantago maritima, .	On the shores.
——— coronopus, .	Dry places.
Atriplex laciniata, . .	Sea shores.
——— angustifolia, .	Shore at Struey.
——— do. (var. erecta),	Slaodridh.
Salicornia herbacea, .	Loch Ranza.
Suæda maritima, . .	On the shores—Loch Ranza, etc.
Salsola Kali, . . .	Sandy shores—frequent.
Rumex conglomeratus, .	
——— pratensis, . .	Lamlash.
Polygonum Raii, . .	Shore at Lamlash, Lag, etc.
———— Convolvulus,	On the shores.
Oxyria reniformis, . .	Mountains.
Empetrum nigrum, .	Lag.
Parietaria officinalis, .	Brodick Castle.
Ulmus montana, . .	
Callitriche pedunculata, .	
Myrica Gale, . . .	Abundant.
Betula alba, . . .	Frequent.
Salix herbacea, . .	Mountain tops.
Corylus Avellana, . .	Abundant.
Populus tremula, . .	Torlin.
Juniperus communis, .	Mountains.
——— nana, . .	Goatfell (?).
Epipactis enisfolia, . .	Slaodridh, Invercloy, Whiting Bay.
Listera ovata, . .	Lamlash.
——— cordata, . .	Cior-Vor, Suithi-Feargus, Glen Cloy.
Orchis mascula, . .	Abundant.
——- maculata, . .	Do.
——- latifolia, . .	Do.
Habenaria bifolia, . .	Do.
——— Chlorantha, .	Do.
——— viridis, . .	Whitefarland, Eais-a-mhor, Loch Ranza.
——— albida, . .	Loch Ranza; and dry heaths, in many places.
Liparis Loeselii (?) .	
Gymnadenia conopsea, .	Abundant in fields and heaths.

Malaxis paludosa, . .	Kildonan.
Allium ursinum, . .	Abundant.
——— * Schoenoprasum,	Glen Shirag.
Agraphis nutans, . .	Abundant.
Juncus triglumis, . .	Goatfell.
——— trifidus, . .	Do.
——— compressus (var. Gerardi), .	Salt marshes.
Juncus glaucus, . .	Torlin.
——— maritimus, . .	Sea shores.
Narthecium ossifragum, .	Abundant.
Alisma Plantago, . .	Brodick, etc.
——— ranunculoides, .	In several places.
Triglochin palustre, .	Frequent.
——— maritimum, .	Salt marshes.
Typha latifolia, . .	Mill-dam, Whiting Bay, etc.
Sparganium ramosum, .	Lamlash.
——— simplex, .	
Potamogeton plantagineus,	Brodick.
——— oblongus, .	Loch-an-Deavie.
Zostera marina, . .	Abundant.
Eriophorum vaginatum, .	Goatfell.
——— polystachion,	
——— angustifolium,	Abundant.
Schoenus nigricans, . .	Sea shores.
Eleocharis palustris, .	Do.
——— do. (var. uniglumis), .	Sands near Kildonan, and at Lag.
——— multicaulis, .	On the coast.
Blysmus rufus, . .	On the sea shore.
——— compressus, .	Do.
Scirpus maritimus, .	Do.
——— pauciflorus, .	Corriegills and Corrie shore.
Carex ampullacea, . .	Near Mauchrie.
——— lævigata, . .	Roadside between Brodick and Lamlash; Loch Ranza, Corriegills.
——— vulpina, . .	Shores.
——— pauciflora, . .	Ascent of Goatfell, and in Glen Rosa.
——— arenaria, . .	Sandy shores.
——— remota, . .	
——— curta, . . .	On the moors.
——— divisa, . . .	Frequent.
——— distans, . .	Imochair.
——— extensa, . .	
——— fulva, . . .	
——— glauca, . . .	
——— muricata, . .	Dippen.

Carex ovalis, . . .	
——- Oederi, . . .	Near Dugarry.
——- paniculata, . .	
——- rigida, . . .	Goatfell.
Rhyncospora alba, . .	North Glen Sannox, etc.
Isolepis Savii, . . .	Corrie.
——— setacea, . .	Frequent.
Avena planiculmis, . .	Said to have been found in Glen Sannox.
Ammophila arundinacea,	On the shores.
Elymus arenarius, . .	Sandy shores.
Briza media (?), . .	
Aira alpina, ...	Goatfell (?).
Brachypodium sylvaticum,	Woods.
Catabrosa aquatica, .	Shore at Kildonan and Lag.
Festuca bromoides, . .	Lamlash.
——— gigantea, . .	Corriegills.
——— vivipara, . .	Goatfell.
Molinia coerulea (var. alpina),	On the mountains.
Poa maritima, . .	On the shores.
Triticum junceum, . .	Shore at Slaodridh.
———-- laxum, . .	Sands at Struey.
Polypodium vulgare, .	Very common.
———— Phegopteris,	Abundant in the woods, and along the sea-cliff; frequent also in sheltered spots in the glens; and sometimes found in rock crevices at the tops of the highest mountains.
———— Dryopteris, .	In the woods, and very abundant in damp sheltered spots in the glens and among the heather; often ascending the highest mountains along with the last species.
Aspidium lobatum, . .	Frequent on cliffs and banks near the sea; sometimes in exposed situations, passing into the form lonchitioides.
——— aculeatum, .	Frequent in hedges and woods.
——— Oreopteris, .	Very abundant in the glens and on the moors, often, as in the lower part of Glen Rosa, covering the whole hill-side with the delicate yellowish-green of its young fronds.

Aspidium spinulosum, .	This variable and perplexing species is very abundant in several varieties, one of which, with an ovate or ovate-lanceolate frond, is frequent on the highest mountains.
——— Filix mas, .	Abundant everywhere.
——— recurvum (Foeni-secii), .	Frequent on the sea-cliff—as at Corrie, North Sannox, Salt Pans, and Whiting Bay; in rocky and wooded ground at the head of Glen Cloy, under the Sheeans.
Cystopteris fragilis, .	Rocks at the head of Glen Cloy, and in Bein-Leister Glen; cave at Blue Rock; and in several other spots among the mountains, as, near the top of Ben-Ghnuis, on the Ceims, etc. It generally, if not always, occurs in the form *dentata*.
Asplenium Ruta-muraria,	Old walls at Brodick Castle, and probably in other places.
——— Trichomanes,	Common on rocks and walls.
——— viride, . .	Abundant on limestone cliffs at the head of Bein-Leister Glen, above Lamlash; on rocks near the head of Glen Cloy, sparingly.
——— marinum, .	Occurring here and there in caves along the old sea-cliff. It is now scarcely to be found on the east and north coasts of the island, but may be obtained in many places on the south and west, as at Benan-head, King's Cove, and Imochair. At Leac-a-Bhreac, near Blackwater-foot, we have gathered fronds eighteen inches long.
——— Adiantum nigrum,	Abundant on banks and rocks.
——— Filix-fœmina (Athyrium Filix-fœmina),	Very abundant and beautiful, and presenting an infinite variety of forms.
Scolopendrium vulgare, .	Hedges, banks, and rocks; often very luxuriant.
Pteris aquilina, . .	Common.

Cryptogramma crispa, .	Among granite blocks at the head of Glen Sannox, under the crags of Cior-Mhor, abundant; also on the pass leading from Glen Rosa into Iorsa, in small quantity, and probably in several other places. One very small plant we found, on one occasion, at the summit of Ben-Ghnuis.
Blechnum boreale, . .	Common.
Botrychium Lunaria, .	Fields near the String Road in Glen Shirag; also on the shore at Invercloy and Corriegills, and in dry pastures in several other places.
Trichomanes radicans, .	Recently discovered by a native botanist on the old sea-cliff, in a single spot where it is undoubtedly wild, not introduced; and and from which it would be very easy to extirpate it.
Hymenophyllum Tunbridgense, . .	On sheltered rocks in several places, as in a plantation near Invercloy, in the wood at Brodick Castle; near Corrie; at Fallen Rocks; and at the entrance of Glen Eaisdale, Whiting Bay. It is remarkable that, in all these localities, it is found growing upon the old sea-cliff; as is also the case where it occurs at Dunoon, Holy Loch, the Kyles of Bute, and the shores of Lochlomond at Tarbet.
——— Wilsoni, . .	Abundant in many spots, as near Corrie; sea coast at Sannox; Glen Cloy; Bein-Leister Glen; Birk Glen, above Invercloy, often growing with the last-mentioned species, but also ascending to the tops of the highest mountains, where, as on Ben-Ghnuis, it covers the most exposed rocks at a height of 2,500 feet, and is with difficulty distinguished from a moss. In general, it prefers damper spots than H. Tunbridgense, and stands exposure much better.

Osmunda regalis, . .	Frequent, on the sea cliff, all round the island, often very luxuriant, attaining the height of 10 or even 12 feet. It is now less abundant than formerly on the east coast, but may be found in profusion near Loch Ranza, and at King's Cove.
Ophioglossum vulgatum, .	In a meadow between Benan-head and Torlin, and probably in other places.
Lycopodium clavatum, .	Abundant on the hills.
——— Selago, .	Moors and mountains frequent.
——— annotinum, .	Said, but in all probability erroneously, to have been found on Goatfell.
——— selaginoides, .	Boggy and springy places on the hills and glens—frequent.
——— alpinum, .	On the mountains.
Equisetum fluviatile (Telmateia), . .	Frequent in damp ground.
——— arvense, . .	Common.
——— sylvaticum, .	In the woods and glens, not unfrequent.
——— limosum, .	Ditches and ponds.
——— palustre, .	Boggy places—frequent.

MARINE ZOOLOGY OF ARRAN

---

106. The *coast-line* of Arran abounds with objects of interest. Nor is this interest confined to geological science. The naturalist—and especially the student of marine zoology—will find here a field for investigation; and even the casual visitor, if endowed with the faculty of inquisitiveness, may obtain instructive and amusing occupation.

Let it be understood that the waters of the sea, generally, teem with animated beings of an endless variety in form, size, and structure, and that myriads of these creatures are the inhabitants of rock-pools, where they share the vicissitudes common to mortality—sporting, at one moment, in the full enjoyment of the pleasures of life, and at another instant either wrangling or wrestling with an antagonist—perhaps for a mouthful of food—or "scudding" with all their might from the open jaws of an approaching and ravenous enemy.

We will suppose that we are now standing at Corriegills, on the sea shore in Brodick bay, about one mile from Invercloy. The rocks are of the red sandstone formation; and the soft material, yielding to the action of ceaseless tides, and to the lashing waves of the winter storm, has been hewn in all directions into fissures and small basins, which, renewed with fresh supplies of water by each returning flood, are the "habitats" of plants and animals. Look into these adjacent pools. They appear to be almost covered with tiny sea-weeds. Yet the jungles of India are not more fully occupied with their appropriate denizens than are these patches of the sea. They abound with microscopic life, generated under the warmth of the sun; and, by stooping down and watching for an instant, you will observe as much bustle, enterprise, and activity, with the usual accompaniments of success and misfortune, as are characteristic of an opulent city. The water at this moment is perfectly still. There is not a breath of air to disturb its surface. But do you notice that ripple? Depend upon it there is some mischief not far off. See! there is a prawn, or something like a prawn, swimming with the greatest eagerness towards a piece of rock almost concealed by marine vegetation. Nay, there is a whole fleet of these creatures, trying to outvie each other in speed, and all shaping their course in the same direction. With the help of our gauze or muslin landing-net we will endeavour to catch a specimen. Skill and patience are required, for they are uncommonly agile—they *leap* with the suddenness and rapidity of a grasshopper, and they are too "knowing" to enter a *bag*, however fine may be its fabric, unless they are enticed or surprised by stratagem. Look how carefully that individual is keeping his *face* towards your net—see how deliberately he "backs astern," with an occasional jump, when you think that he cannot possibly escape:—but now, keep your hand steady; he will approach; he will inspect your net with all imaginable curiosity. Observe how skilfully he uses his "horns," or antennæ—exquisitely formed instruments, half as long again as his entire body, and which, as with the insect tribes, are delicate organs of inquiry. He will creep round the net and touch every part within his reach. But stop—here is a chance of catching him—run him gently back into that nook in his immediate rear, dip your net under him, and then, if you are smart in raising your hand, he will be your prisoner. Bring him on shore, and, inverting the net into a glass vessel of sea water, we will be able to examine him at leisure. But what has become of his companions? Whither were they bound? And on what errand were they so swift? It is a case of piracy. These crustaceans, which we saw swimming in such haste, are a fleet of sea-robbers; they are at this moment plundering a colony of helpless zoophytes, and, as often happens under analogous circumstances, they are quarrelling among themselves. Look, first, at these two warlike specimens of their race engaged in combat—each endeavouring to get at the prey. They do not appear to use their eyes, although these organs are well developed; but, feeling about on all sides with their antennæ, they no sooner *touch* each other than the strongest of the two rushes forward, assaults, and beats back the weaker, which, after being foiled in repeated attempts to hold his position and obtain a share of the spoil, retires to another portion of the invaded colony. Now observe their thieving propensities. Here is a group of sea anemones, just below the surface of the pool, curiously fixed with their base in a chink of the sandstone rock, whilst their upper or anterior parts, fringed with tentacula, are exposed, like the sunflower when its petals are expanded, to the cheering rays of light. These are at present undisturbed. Here is another group; and of these several have made prize of some food, brought within their reach by the graceful vibrations of their tentacula. The crustaceans—whether they *scented* the dainty morsels or not, we cannot determine—have hurried to the spot in order to appropriate as plunder what has been legitimately obtained by others. Each prawn is helping himself, drawing the coveted bits from the grasp of the helpless and disappointed zoophytes: and see how he scampers off with both hands full to a place of retreat. But look once more. You observe an anemone in the act of swallowing a piece of decayed shell-fish—his mouth opens wider and wider—and now the morsel has disappeared. Here is a prawn just arrived—almost too late—he feels with his finger and thumb about the lips of the closed mouth—there is apparently nothing for him; yet he seems to know what he is about;—and true enough; for see how artistically, how professionally he introduces his hand into the mouth of the anemone! He will not be defeated. Nor has he any compunction. He thrusts his hand down, and down, and yet deeper still, into the very *stomach* of the animal. He has actually got possession of every particle that had been swallowed; and now he is off with all possible speed lest his ill-gotten booty should be snatched from him by a prawn more valiant than himself.

This colony of plundered zoophytes is formed by a species of anemone common only on certain parts of the coast—it is the *Actinia bellis*, described by naturalists. It sometimes escapes detection from the fact, that almost the whole body is often concealed within the chink of a rock, whilst the expanded disc and the surrounding tentacula are partially obscured by the shade of adjacent sea weeds. It assumes a variety of shapes; it can change its locality at pleasure; and, having apparently an aversion to the solitary life, it is usually found in clusters of perhaps a dozen or more individuals huddled together in close compact. The species can scarcely be mistaken. In the same pool are other species of the same genus. The most abundant of all the anemones, *Actinia mesembryanthemum*, is seen on all sides. It inhabits nearly every place at the sea side, adhering to stones or to the sides of piers, and is easily recognised, either as a gelatinous rounded body, not unlike a ripe cherry, when closed during the recess of the tide, or, when in the full enjoyment of the returning waters, it exhibits the form of the flower from which its name has been derived. A careful examination of some of the deeper pools in this neighbourhood will probably expose to view very fine specimens of the larger species—*Actinia coriacea* and *Actinia crassicornis*. The former of these is abundant on many parts of the shore, sometimes buried in the sand, and gregarious, or hidden under fragments of rock between the tide marks, or in the possession of a luxuriant pool. It may be known by the following characters:—Body conoid, or wider at the base than above; skin opaque, coriaceous, or leathery, covered with warts, variously coloured, but usually dull red; tentacula numerous, in three or four series, and, when expanded, marked with rings. The other species, *A. crassicornis*, although commonly found attached to shells or stones in deep water, may often be obtained in rock crevices, at low water during the spring tides, or éven in the large pools. It may be procured at Corriegills, or among the rocks at the south-east side of Holy Island. It resembles *A. coriacea* in shape; but it is a finer species; it grows to a larger size, it is less coriaceous, and more vivid in its colours. It is a handsome object for the aqua-vivarium, where, after a short period, it will be observed that the body becomes filled with water, so as to be greatly enlarged and diaphanous, or almost transparent; the tentacula, at the same time, are elongated, increase their diameter, and exhibit their tubular form. This diaphanous appearance is perhaps the most striking character by which it is distinguished from *A. coriacea.* It will add considerably to the interest of the vivarium—provided the tank can afford accommodation—if three or four individuals of this fine species are placed artistically among the rock-work; for, by a careful selection, specimens may be introduced of variegated hues, some of cream colour, some of scarlet, some of saffron, with orange stripes, or marbled with red and white. There are, indeed, certain people who entertain *other ideas* as to the proper destination of *Actinia crassicornis.* "For, of all kinds of sea anemones," says Dicquemare,* "I would prefer this for the *table;*—being boiled some time in sea water, they acquire a firm and palatable consistence, and may then be eaten with any kind of sauce. They are of an inviting appearance, of a light shivering texture, and of a soft white and reddish hue. Their smell is not unlike that of a warm crab or lobster." But whatever importance the *epicure* may attach to the above information, the sea-side visitor will do well to watch the manœuvres by which the anemones are themselves enabled to gratify their tastes by a variety of dishes. It is most amusing to observe the apparent cleverness—the agility and artifice exercised by these curious creatures for the capture of their food. "On one occasion," says Mr Couch, "while watching a specimen (*A. coriacea*) that was covered merely by a rim of water, a bee, wandering near, darted through the

* Johnston's British Zoophytes, vol. i., p. 227.

water to the mouth of the animal, evidently mistaking the creature for a flower, and, though it struggled a great deal to get free, was retained until it was drowned, and was then swallowed."* They are at once the most abstemious and the most gormandizing of animated beings. They will live without food for upwards of a year, and yet they may be seen at all hours, and every day, angling, as it were, with their tentacula, and catching crabs, prawns, limpets, periwinkles, dog-whelks, small fish, and, in short, whatever, in the shape of fish, flesh, or fowl, is brought within their reach. They retain their food for ten or twelve hours, and then eject from the stomach the well-picked bones—the emptied shells of the crustaceans and mollusca. Occasionally a bone will stick in the throat. "I had once brought me a specimen of *A. crassicornis* that might have been originally two inches in diameter, and that had somehow contrived to swallow a valve of *Pecten maximus* of the size of an ordinary saucer. The shell, fixed within the stomach, was so placed as to divide it completely into two halves, so that the body, stretched tensely over, had become thin and flattened like a pancake. All communication between the inferior portion of the stomach and the mouth was of course prevented, yet instead of emaciating and dying of an atrophy, the animal had availed itself of what undoubtedly had been a very untoward accident, to increase its enjoyments and its chances of double fare. A new mouth, furnished with two rows of numerous tentacula, was opened up on what had been the base, and led to the under stomach:—the individual had indeed become a sort of Siamese twin, but with greater intimacy and extent in its unions."† Another remarkable character in the natural history of the *Actiniæ* is, that, though impatient of ill-treatment under certain conditions, as when torn abruptly from their attachment to the rock, or when confined in water not sufficiently pure, they are almost indestructible by the usual methods of destruction. "They may be immersed in water, hot enough," says Dr Johnston, "to

* Johnston's British Zoophytes, vol. i., p. 225. † Ibid, p. 235.

blister their skin, or frozen in a mass of ice and again thawed; and they may be placed within the exhausted receiver of the air-pump, without being deprived of life, or disabled from resuming their usual functions when placed in a favourable situation. If the tentacula are clipped off they soon begin to bud anew, and if again cut away they grow again." * The finer specimens of these two species, as found in this locality, will occasionally measure from four to six inches across.

107. With the exception of the above examples, the coast line of Arran is not prolific in a variety of zoophytes. A few other forms may be procured in deep water by means of the naturalist's dredge. There are, however, two more species that will possibly attract the attention of the inquisitive rambler along the shore: they belong to genera closely allied to *Actiniæ*, namely, *Anthea cereus*, and *Adamsia palliata*. The former frequently chooses for its position the leaves of the grass wrack, or sea grass (*Zostera marina*), extensive beds of which constitute submarine meadows in sandy districts. This wrack (including the other species, *Z. nana*) is the only instance of a British flowering plant that lives in the sea. Now, wherever there is a bed of this plant, the student of marine zoology (and also the algologist, or marine botanist) may pause and examine; and, if he does not object to *wade* knee deep, at ebb-tide, he will almost to a certainty be repaid for his trouble. Large numbers of creatures revel and thrive in this forest—fish, crabs, shell-fish, annelids or sea-worms, and zoophytes; and here, dependent from the long, riband-like leaves of the zostera, you may often discover *Anthea cereus* with his lengthened tentacula, on the look-out for his prey. The following characters will be sufficient for identification:—body cylindrical, smooth, adhering by a broad base; tentacles numerous, longer than the body, and, unlike those of the *Actiniæ*, these arms cannot be retracted, or are scarcely retractible, into the body of the animal. It attains in this neighbourhood the size of about three inches, measured from

* Johnston's British Zoophytes, vol. i., p. 239.

the base to the oral disc, or mouth; and it is of a light brown, or dull ash colour. Beds of the *Zostera marina* occur at Lamlash, on the sandy shore opposite the village; also, on Holy Island, not far from St Molios' cave; but, even where the sea wrack does not grow, *Anthea cereus* may be found attached to one of the commonest of our sea-weeds—*Fucus serratus.* The other species—*Adamsia palliata*—will not so readily be discovered by the casual visitor, for it prefers the deep water. The dredge, when let down to ten or twenty fathoms, will often bring up a dozen specimens at a time. This zoophyte may be grouped amongst the most curious of creatures. It will be sure, when once seen, to arrest attention. Its *habitat,* or dwelling-place, is the exterior wall of a deserted shell, as, for example, the dead shell of a *buccinum,* or whelk, or of a *trochus,* or *fusus,* over the greater part of which the animal is extended as a flattened mass about three-tenths of an inch in thickness, varying from a light brown to cream colour, and having the whole surface of its body streaked and spotted. The situation of the oral disc is distinguished by three or four rows of tentacula, short and white, forming an oval margin round the mouth. A thin substance, like horn, will be observed covering part of the empty shell, to which horny material the body of the *Adamsia* is attached; and, when excited, apparently by pleasure, as with feeding, it emits long filaments, like white threads or delicate silk cords. But a singular circumstance has yet to be mentioned:—The dead shell, over which this zoophyte is spread, is generally inhabited by a hermit crab, and always, as is supposed, by the same species, namely, *Pagurus Prideauxii.* It is exceedingly curious to watch how advantageously to both parties the arrangement works. Of course the Adamsia, without any fatigue or effort, is carried by the roaming propensities of the crab over a large district; and, in this way, he commands an extensive market for the acquisition of food; whereas the *Actiniæ,* being fixed to rocks or half buried in the sands, must either undertake a slow and wearisome journey, by their own unassisted labour, or be satisfied with the supplies brought by the wind or tide within reach of their feelers. The hermit, on the other hand, is also recompensed; for, as the writer of these notices has frequently observed, the palatable morsels secured by the tentacles of the zoophyte are instantly seized by the claws of his crustacean companion, and, without any apparent apology or subsequent remorse, are partly appropriated to his own immediate use. "In all likelihood," says the late Rev. Dr Landsborough, "they in various ways aid each other. The hermit has strong claws, and while he is feasting on the prey he has caught, many spare crumbs may fall to the share of his gentle-looking companion. But soft and gentle-looking though the *Actinia* be, she has a hundred hands, and woe to the wandering wight who comes within the reach of one of them; for all the others are instantly brought to its aid, and the hermit may soon find that he is more than compensated for the crumbs that fell from his own booty." * Specimens of this curious and beautiful zoophyte may occasionally be procured by a search in the zostera beds at Lamlash, or in other places at low water. Dr Landsborough first saw it at the mouth of the Glen Rosa burn, in Brodick Bay.

108. We must now inquire about the prawn captured at Corriegills. Look at it through the sides of the glass vessel. The creature is almost transparent. This is *Palæmon Squilla*; it is smaller than the common prawn, *Palæmon serratus*, and differs in a few other points from that species. It is readily found in the rock pools on Holy Island, on both sides of Brodick Bay, and indeed almost everywhere on the Arran shore, and on the Ayrshire coast. It is an interesting object for the vivarium, both on account of its general appearance, and from its activity in plundering the sea anemones. Other crustaceans abound in this district. The numerous rocks are places of retreat and shelter for the common edible crab, *Cancer Pagurus*, which, in moderate size, is caught in the creels by local fishermen, whilst the younger individuals of this species

* Landsborough Popular History of British Zoophytes, p. 230.

may be recognised scampering about the shore in search of food, or hiding themselves under stones, in the enjoyment of a siesta. The common shore or harbour crab, *Carcinus Mœnas*, inhabits the same localities, and may be seen anywhere, or everywhere, in pools, under sea-weeds, beneath rocks, or in the sands. Turn over the stones, and another species will invite inspection—*Porcellana platycheles*—distinguished at once from every other crab by the breadth of its hands. But, of all the crustaceans, the hermit crab, already mentioned in connection with an associated zoophyte, seems to afford the most amusement to the sea-side rambler. There are several species of this singular crab inhabiting deep water; but the most common kind, *Pagurus Bernhardus*, may be procured in any quantity by strolling along the shore. The peculiarities of these hermits are of no mean order. Each individual resides, hermit-like, solitary in a shell, which, either by fair or foul means, is obtained and appropriated for self-protection. Nature has been considerate in supplying a shield (the carapace) that entirely defends every species of crab, except the genus now under our notice, namely, *Pagurus*, which, whilst tolerably safe as to the head and thorax, is in constant danger from the utterly unprotected state of the posterior portions. The abdomen and tail are without the usual covering of a coat of lime; they have no *shell* in these parts, and, consequently, at any moment they might be at the mercy of a hungry neighbour or spiteful antagonist. But here, as elsewhere in the wonderful economy of Nature, the animal finds *compensation* for the deficiency, either in the use of its instinct, or rather, perhaps, in the exercise of *good sense*, by which a remedy is provided equal to its necessities. Its usual habit is to take possession of the empty shell of a dead shell-fish—one of the univalves, such as the common whelk—and, introducing the whole of its soft parts into this cherished prize, it hooks its tail round the innermost whorl of the shell, and keeping both eyes and claws ready for action, it bids a hearty defiance to every opponent. You may seize the exposed claws and tear the hermit into pieces; but, so apprehensive are they of danger, and so tenaciously do they cling to their support, you will seldom either persuade or force the creatures to come out of their abodes. It is equally curious to observe the cleverness shown when required to meet another difficulty. A *young* hermit occupies a *small* house. This is a necessary conclusion, inasmuch as the shell has to be dragged about by its occupant. But juvenile hermit crabs, like other young creatures, increase in size, and, as they grow, they find that the abodes selected in their youth have become inconveniently small; for, of course, the dead shells cannot be enlarged for their accommodation. This embarrassment is easily obviated. It is merely a matter of change of residence. You will see the hermit—if you watch on the sea shore, and happen to be in the right place at the right time—bring himself into close proximity to a shell, empty, and larger than the one then in his possession. His next step is to take an accurate survey on all sides, to see that the coast is clear, and to make sure that he can move from one house to the other without the risk of a *disadvantageous* encounter with an enemy. It is the work of a moment, provided that the new house proves to be more commodious than the old one; but it sometimes happens that the crab is mistaken in his calculation, and, in this case, it is most amusing to watch our friend running about, half naked, and half frightened, among a group of deserted shells, poking its tail first into one and then into another, twisting itself in various directions, and assuming different attitudes, in order to test the suitability of the proposed arrangement; and, finally, when fitted to its entire satisfaction, it walks off, evidently pleased with its success, and generously leaves the old shell for the accommodation of any future house hunter less corpulent than itself. Not unfrequently two hermits will meet and dispute the possession of a vacant shell. Arbitration is never sought in these cases. It is simply a contest between the two for mastery—a trial of physical power—and, where the combatants are well matched, the conflict is generally of a fatal character; for the possession of a convenient house is considered of such vital importance that the one or the other is determined to obtain the coveted shell, and, at all times pugnacious, each is resolved to die rather than yield ingloriously to its opponent. This sea-fight often occurs. The slain are not a few. Sometimes a less serious encounter may be witnessed, as when *might* exercises its too frequent tyranny over *right*. The following incident affords a good illustration: Two hermits met, accidentally, it may be supposed, within the vivarium on Holy Island. The one was somewhat stouter than the other; and the weaker brother, appearing at first sight to possess the larger house, received an immediate summons to turn out and vacate his premises. The summons was disputed, a furious combat commenced, and, after a variety of manœuvres, with thrusts and counter thrusts, the stronger, with a dexterous use of the hands, seized the little fellow by both his wrists, and endeavoured to *shake* him out, despite the opposition that was offered. The shells rattled against the glass wall. It was a fierce engagement. At last the older hermit, coming partly out of his shell, seemingly for the purpose of obtaining a better purchase and a greater command of his strength, pulled, and pulled, and pulled again—still drawing himself farther out of his shell, as if he felt that the work was desperately tough—until the youngster, giving way, stood, unhoused, discomfited, and embarrassed, to abide the good pleasure of his master. The next scene was amusing enough. The conqueror came entirely out of his house, and —retaining a firm grasp of his victim to prevent him from bolting off with either of the two shells now empty—backed himself into the new abode, just to inquire if it would really prove a more convenient residence than his own, but, disappointed in his anticipations, he returned to his former shell, released his young friend without further detention or injury, and, apparently with a polite "good morning," they separated as if nothing discourteous, or at least unusual, had occurred.

109. The star-fishes will also afford interest during a ramble

along the shore. The most common species, *Uraster rubens*, is really too common to be further noticed. Some kinds, as *Palmipes membranaceus*, *Cribella rosea*, *Asterias aurantiaca*, and others, can be procured only by the dredge; but several of the deep water species may occasionally be found by wading through the zostera beds at Lamlash, or by examining the pools and rock crevices, especially after a storm, on Holy Island. In this way, *Luidia fragilissima*, *Uraster glacialis*, *Cribella oculata*, *Goniaster Templetoni*, *Solaster papposa*, with some of the species of *Ophiocoma* and *Ophiura*, may be picked up without much trouble. In a few places, the smallest of British star-fishes, *Asterina gibbosa*, may be obtained; but good eyes and some patience are required. Try the pools at the north end of Holy Isle; and also at Clachland point, opposite. Turn aside the sea-weed; and it may be, after considerable search and disappointment—for this species is by no means abundant—you will see the little creature, grayish in colour, and gibbose and angulated in form, adhering to the side of the rock. In Lamlash bay, *Uraster glacialis* is plentiful. It may often be observed, on a calm day, when the water is as clear as crystal, directing its course among the algæ at the depth of 10 or 15 feet—the creature itself, perhaps, two feet in length—and, on nearer inspection, you will distinguish it at once from *Uraster rubens* by the spines situated on the back of each ray. Care is needed in the preservation of this species, for it has the knack of dislocating its arms under your most cautious treatment. But the star-fish that has gained most notoriety from the facility with which it can, apparently at pleasure, dismember itself—nay, break itself into any number of fragments—and, vexatiously enough, at the very moment when you think the animal is deceased, and preserved uninjured, is *Luidia fragilissima*. The subjoined description is from the pen of the late Professor Edward Forbes:—"The first time I ever took one of these creatures I succeeded in getting it into the boat entire. Never having seen one before, and quite unconscious of its suicidal powers, I spread it out on a rowing bench, the better to admire its form and colours. On attempting to remove it for preservation, to my horror and disappointment I found only an assemblage of rejected members. My conservative endeavours were all neutralised by its destructive exertions; and it is now badly represented in my cabinet by an armless disk and a diskless arm. Next time I went to dredge on the same spot, determined not to be cheated out of a specimen in such a way a second time, I brought with me a bucket of cold fresh water, to which article star-fishes have a great antipathy. As I expected, a *Luidia* came up in the dredge—a most gorgeous specimen. As it does not generally break up before it is raised above the surface of the sea, cautiously and anxiously I sunk my bucket to a level with the dredge's mouth, and proceeded in the most gentle manner to introduce *Luidia* to the purer element. Whether the cold air was too much for him, or the sight of the bucket too terrific, I know not, but in a moment he proceeded to dissolve his corporation, and at every mesh of the dredge his fragments were seen escaping. In despair I grasped at the largest, and brought up the extremity of an arm with its terminating eye, the spinous eyelid of which opened and closed with something exceedingly like a wink of derision."*

Experience has since shewn that the means by which either *Luidia* or *Uraster glacialis* can be secured for cabinet specimens is, not to destroy them violently by plunging them into fresh water, but to let them die by a more quiet process, viz., by leaving them in a vessel of sea water until life is exhausted. Under these circumstances they are not prone to break, and thereby disappoint their captors.† Luidia will be recognised by its light orange or buff colour; its rays are smooth on the back, and provided with spines on their margins; specimens measuring a foot may be picked up on the shore in Lamlash bay, and individuals of twice that size are procurable by the

* British Star-fishes, page 138.

† Dr Carpenter has found that, by placing Luidia in a pan of *Glycerine* the creature dies at once, and without disfiguring itself.

dredge, or may sometimes be hooked up in shallower water with a common rake. There is another singular creature of great interest to the naturalist—it is *Comatula rosacea*, or the rosy-feathered star-fish—a beautiful and elegant example of the radiated form of animated beings. It is found abundantly in Lamlash bay, near the pier on Holy Island, and in other parts of the locality; but the dredge is required, as this species inhabits deep water—that is, depths of about ten fathoms. At some seasons almost any number may be obtained. They are brought up attached to the large sea-weed, *Laminaria saccharina*, from which they must be removed with care, as they are exceedingly brittle, and like the *Luidiæ*, can break themselves into fragments with astonishing and unpleasant speed. The best way, perhaps, to prepare dry specimens for the cabinet is to treat them as the marine botanist treats the more delicate algæ—spread them on drawing paper, place over them a piece of smooth linen, and let them dry between folds of blotting paper, under slight pressure; but, in the first instance, they must be allowed quietly to die in sea water, or more rapidly in a solution of alum—for, if immersed in *pure fresh* water, they will lose their beauty, by the removal of their colouring matter. In order to appreciate the exquisite form and elegant movements of this star-fish, the creature must be seen and watched in a vivarium, where it will voluntarily fix itself to the sea-weed or to a piece of rock, and, by graceful undulations of its arms, will be certain to command admiration. The picture will be greatly enhanced if several individuals of various colours—orange, purple, crimson—are introduced, and judiciously dispersed.

110. Before we leave this division of the subject, it will suffice to make brief mention of the well-known sea hedgehog, *Echinus sphæra*, common in most parts of the Clyde, and of which fine specimens are found clinging to the rocks on Holy Island, and at Clachland point. This curious creature should be carefully examined by the young student of marine zoology. It is allied to the star-fishes; for although spherical in form, the

radiated structure is readily perceived. In short, it is a star-fish, with the spaces between its rays filled up by plates of carbonate of lime—the rays themselves consisting of the same material—the whole exterior being bent over into a hollow ball, and armed, hedgehog-like, or like the star-fish *Uraster glacialis*, with numerous sharp spines. The viscera or digestive organs of the echinus are contained within the ball; and its mouth is provided with a beautiful piece of mechanism, worthy of examination, and designed for crushing the shells of molluscous and crustaceous animals on which it feeds. Its mode of progression, by means of its spines and suckers, is both interesting and wonderful. Specimens are frequently found of the richest crimson or purple. Another species of this genus, *Echinus miliaris*, is also common in the pools. It is more diminutive than the former. The dredge will generally be needed to obtain living specimens of the other forms of these echinodermatous or radiated creatures—such as *Spatangus purpureus*, *Echinocyamus pusillus*, and *Amphidotus cordatus*, or *the common Heart Urchin*, of which the dead and empty shells, with their spines rubbed off, may frequently be noticed washed up and left upon the sands.

111. The study of another important division of animated nature—the mollusca—has been a favourite pursuit of scientific men and of amateur observers. To this division the marine shells belong. Here the conchologist finds his delight in the examination and arrangement of the materials of his department—admiring the endless variety of form, and colour, and sculpture; whilst the zoologist is laboriously occupied in determining the anatomical structure of the *inhabitants* of the shells—tracing the peculiarities of the different genera and species.

The waters around Arran produce a considerable number of shells; but, with few exceptions, the species are common to all parts of the Clyde. The *dredge* is required to procure specimens of interest; and for the information of naturalists unacquainted with the district, it may be stated, that the most profitable dredging ground in Lamlash bay, lies between Hamilton Rock, near Clachland point on the Arran shore, and the north and north-east sides of Holy Island, extending the whole way across, and in depths from fifteen to forty fathoms. A full list of these species will be given at the end of these notices; but the following genera, as being found in this particular locality, may be here mentioned:—*Aporrhais*, *Artemis*, *Astarte*, *Cardium*, *Cerithium*, *Ghemnitzia*, *Circe*, *Corbula*, *Crania*, *Cylichna*, *Cypræa*, *Cyprina*, *Dentalium*, *Emarginula*, *Eulima*, *Fissurella*, *Fusus*, *Kellia*, *Leda*, *Lima*, *Lyonsia*, *Mangelia*, *Modiola*, *Montacuta*, *Nassa*, *Natica*, *Nucula*, *Odostomia*, *Pecten* (including *P. striatus* and *P. tigrinus*), *Pectunculus*, *Philine*, *Pileopsis*, *Pilidum*, *Psammobia*, *Puncturella*, *Rissoa*, *Scalaria*, *Scaphander*, *Tellina*, *Terebratula*, *Thracia*, *Trichotropis*, *Trochus*, *Turritella*, and *Venus*. Several of these genera occur also in other places in Lamlash bay, and round the coast. On the other hand, Brodick bay, Whiting bay, and the vicinity of Pladda, have hitherto proved to be exceedingly unprofitable. Loch Ranza, notwithstanding the apparent advantage of its position, did not yield anything, after several hours of active dredging in all parts of the bay, except the commonest of the scallops, *Pecten opercularis*—an excellent bait for fishermen, and which may be procured here in any quantity. The strait which separates Arran from Argyleshire, known as Kilbrannan Sound, will probably yield richer results than other parts of the Clyde; but these waters have not yet been sufficiently examined to warrant more than a conjectural opinion. The south end, and the south-east side, of Holy Island are also unprofitable.

It must not be expected that the rarer molluscous animals or shells may be obtained from the pools, or from the sands or rocks forming the coast-line. Yet an attentive investigation will not be without its reward. In the vicinity of Clachland point, where the excavated sandstone affords numerous places of retreat, and where each returning tide supplies the wants of the tenants, a variety of creatures may be seen grouped together in quest of the enjoyments of life. Let the crevices be carefully searched—turn over the fronds of the sea-weeds—capsize the loose stones—look under the ledges of the rocks—select a calm day, during the spring tide, when the water has ebbed to the lowest—and there will not be any need, at least for the young student, to complain of the result. A curious mollusc inhabits this station—*Aplysia hybrida*, or the sea hare, which will be recognised, whilst in a state of activity, by the peculiar shape of its antennæ; for these, when expanded, are something like the erect ears of the common hare. Another mollusc, found upon the shore, and not without pretensions to respect, as well from its lovely orange colour, as from the gracefulness with which it moves, is the *Pleurobranchus*, which, being nocturnal in its habits, may be caught napping at the period most convenient to its captor. It is a good object for the vivarium, where, lying concealed during the day, it will be observed, at midnight, traversing the tank, like a solitary watchman on his rounds; but, unlike the guardian of our streets, its purpose is not to defend property, but to appropriate to its own use, without fear of detection, whatever it may regard as palatable to its taste, or requisite for its necessities. Both the *Pleurobranchus* and the *Aplysia* may be procured from pools at the north end of Holy Island. At Lamlash, in the neighbourhood of the old quay, the blocks of sandstone, which lie scattered in all directions, contain, here and there, good specimens of one of the *borers*—a division of the mollusca, whose habit is to penetrate either sand, as the common cockle, *Cardium edule*—or wood, as the different species of *Teredo*—or stone, as the genus *Pholas*, a bivalve, of which the shell in some species is extremely fragile, and although wonderfully well adapted to its work as an excavator of solid rocks, requires the utmost caution when being handled in our collections. The species that bores through the sandstones at Lamlash, is *Pholas crispata*. The wood-borer, *Teredo navalis*, is common in many parts of the Clyde, as may be proved by witnessing its destructive operations at Ardrossan and Fairlie on the Ayrshire side, where also its ally, *Xylophaga dorsalis*, has done its full share of mischief. Another species, less common, and of larger growth—formerly unknown, as is supposed, higher up the Clyde than Port-Patrick—namely, *Teredo Norvegica*, introduced itself into Lamlash bay, and, during the short period of about seven years, it had almost demolished the massive supports of the pier—a commodious and substantial landing-place, erected, as the owner undoubtedly thought, for the convenience of himself and friends. The *Teredo*, however, commenced and continued its operations unnoticed—not one pile of the water-covered timber escaped—the whole was pierced in lengths varying from a few inches to about two feet, when, in the hurricane of February, 1856, a vessel was driven upon the spot, and almost the entire structure was swept away.* On an examination of the fragments thrown upon the island, several specimens of the calcareous tube, formed by the animal in the course of its progress, and a few of the valves were secured for private cabinets; and pieces of the bored timber were given to the museums of Glasgow College and of the Andersonian University.

The most abundant species of *Trochus* on these western shores is *T. umbilicatus*, which, with *T. cinerarius* and *T. tumidus*, may be procured in any quantity on Holy Island. Here also the following shells are not uncommon:—*Acmæa testudinalis*, *Kellia rubra*—a minute bivalve attached to the dark low growing plant, *Lichina pygmæa*, which overspreads the rocks near the sea;—*Patella athletica*, *P. pellucida*, *Tapes decussata*, with different species of *Littorina*, and other equally common and widely diffused forms. In the sands, at Lamlash, at low water, *Mactra solida*, and *M. subtruncata*, are plentiful. But, leaving the coast line, and letting the dredge drop in about ten fathoms, at the north end of Holy Island, or between the north point of the island and the pier, a peculiarly interesting shell may be collected in large quantities—*Lima hians*, the

* This pier was built by the late Mr Oswald, Member for the City of Glasgow.

inhabitant of which constructs and occupies a *nest.* The shell is of delicate texture, and, when deprived of extraneous matter by careful washing, is entirely white, and, being graceful in form, is an acquisition to an ornamental cabinet. It is desirable, however, to preserve specimens of the nest as well as of the shell, and to allow both to remain, as far as possible, in their natural position. The nest is formed of materials collected at the bottom of the sea—either pebbles, or broken shells, or both shells and pebbles—and these are brought into a mass and bound together by a glutinous thread secreted by the animal. Sometimes the *Lima* is solitary—having built a house simply for itself; but, generally in Lamlash bay, the dredge brings up large patches of this compacted debris, in which will be found a colony living in apparent comfort, security, and friendship—each individual, however, having a separate and snug berth. A supply of specimens may be secured in a few moments. The *animal* is also a beautiful object in the water—its numerous tentacula, which extend beyond the limits of its shell, are of a fine orange colour, and, being a bivalve, it moves from place to place with ease and agility, by the rapid opening and closing of its valves. Although rare in many localities, *Lima hians* has an extensive range in the Clyde—the nests being found off the coast of Islay, between Largs and the Cumbraes, and on the Argyleshire side, as high up as Hunter's Quay, Dunoon. But the station where it is really abundant is Lamlash bay, where also the other species, *L. loscombii,* and *L. subauriculata,* may be procured.

112. Although many objects of interest have now been mentioned; there remain a considerable number of living creatures, more or less wonderful both in structure and in habit, which, from various causes, have not yet been sufficiently examined on this coast; and, consequently, any information respecting them must necessarily be meagre. Among the *Annelids,* or sea-worms, the following genera are known to belong to Lamlash bay:—*Aphrodita, Arenicola, Eunice, Nereis,*

Pectinaria, Serpula, Spirorbis, Terebella. The young student will be usefully employed in the search and study of these curious forms of life, for, notwithstanding that they take no higher rank than worms, they exhibit, in some species, characteristics so peculiar, and colours so gorgeous, as to attract even the popular eye and to excite unsparing admiration. Who is not familiar with the *Serpulæ,* whose heads are crowned with radiating threads of varied hue? These singular tufts, which the animal protrudes for health and pleasure, and which, with the rapidity of thought, are withdrawn into its calcareous tube on the first symptom of alarm, are its gills, or organs of respiration. Let the shadow of your hand pass near the side of a glass vessel, in which a living specimen is contained, and instantly the head starts back into concealment; but watch, and in a few moments its brilliant coronet will reappear. The *Pectinaria* resides in a house made of the finest sand, cemented in the form of a tapering tube; and its gills, in shape like a minute comb, are as bright as burnished gold. In the *Terebellæ,* inhabitants of mud, the numerous and long, worm-like tentacles present the appearance of so many separate *Annelids* entwined around their common prey. The genus *Eunice* occupies a tube composed of a substance not unlike thin horn, or the slender quill of a bird. The spiral white spots, frequently spread over tangle or other large sea-weeds, are examples of *Spirorbis.* The sea-mouse, *Aphrodita aculeata,* obtained with the dredge, will easily be recognised by the metallic lustre of its long bristles, which, partially covering the animal, give out the colours of the rainbow. Some of the marine worms are remarkable from the enormous length to which their thread-like bodies are extended. Another class of widely diffused beings, not yet sufficiently investigated on the Arran shores, and which, like the *Annelids,* are worthy of a more honourable name than is assigned to them in popular language, are the *Sea-slugs,* known in scientific phraseology, as the *Nudibranchiate mollusca.* These are within the reach of every observer—for, at low water, they may be seen reposing under loose stones or adhering to *Algæ.* Some of the species, especially of the genus *Eolis,* are really charming objects—elegant in form, and beauteous in colour. They cannot be mistaken. Place a specimen in a tumbler of sea water, it will unfold itself, and its gills, differently situated in different species, will be seen expanding into full operation as the little creature pursues its journey round the sides of the vessel. It enjoys the learned appellation of *Nudibranch,* because its *branchiæ,* or breathing organs, are *naked,* or exterior to the body of the animal. The genera *Doris, Eolis, Goniodoris, Lomanotus, Polycera,* and *Triopa,* have been found in the bays of Brodick and Lamlash; but, if an active search be made, the number of ascertained species will doubtlesss be greatly augmented, and the labourers in this department will be amply rewarded. Nor must we altogether overlook, as among the more remarkable forms, the *Acalepha* to which the *Medusæ* and other *Jelly-fish* or *Sea-nettles* belong. Some of the larger and coarser species are familiar to sea-bathers by the stinging qualities of their tentacula. But there are other species as harmless as they are lovely. On a calm summer day, when there is not a ripple to disturb the sea, these exquisitely formed creatures may be witnessed in hundreds, like a vast fleet of fairy ships, lying upon the surface. Such are the *Beroe* and the *Cydippe.* Let specimens, captured carefully in a gauze net, be transferred to a vessel of water, and, low as they are in organization, they will not be dismissed without commanding surprise at the delicacy of their structure, and at the facility with which they traverse their allotted space.

113. It remains only to add, after the above general view of invertebrate life, that in the various bays and inlets around Arran, almost every kind of *fish* common to our northern seas may be procured. It would be superflous to name the species familiar to every person. But we must not omit to mention that the Lancelot—*Amphioxus lanceolatus*—has been dredged at the north end of Holy Island, where it seems to be restricted to a gravel bottom in depths of about ten or fifteen fathoms.

Until recently this fish was regarded as extremely rare. It is now, however, known to be more common. Several specimens have been captured near Millport, in the island of Cumbrae, as well as in Lamlash Bay; and probably it will be discovered to have a wider range than at one time was expected. The interest attached to it arises from its anatomical and physiological peculiarities. Its spine is a cartilaginous, thread-like column, without joints; it has no ribs, no pectoral or ventral fins; and, in short, although claiming rank with vertebrate animals, the skeleton is rudimentary, and the brain absent. The *Amphioxus* is an excellent illustration of the law that, even when there is the greatest departure from uniformity, the *typical characters* are rigidly preserved in the development of creation.

114. In concluding our notices of the marine zoology of Arran, the subjoined tables may advantageously be added for the information of naturalists not practically acquainted with the Fauna of the Clyde. The species marked with an *asterisk* were dredged by the late Rev. Dr. Landsborough and Major Martin; and the other species were obtained by Dr Greville and myself in our examination of this part of the coast. The list, so far as completed without the assistance of the former gentlemen, appears in the Annual Report of the British Association for the Advancement of Science for the year 1856. Dr Landsborough and Major Martin were indefatigable in their investigation of these localities. There yet remains, however, much work to be accomplished by future inquirers—especially among some of the remarkable groups, as the Cirripeda Annelida, Acalepha, Zoophyta, and Poriphora—as also among the Polyzoa and microscopic forms included in the Infusiora and Rhizopoda.

NOTE BY WILLIAM B. CARPENTER, M.D., F.R.S., &c.

I am not able to add anything to Dr Miles' list except on one point, which is a very curious one. There are *two* species of *Comatula* (Art. 109) in Lamlash bay; one of them, the commonest, being that described by Edward Forbes as *Comatula rosacea;* whilst the other, rare there, but *the* species of the south of England, is that described by Miller in his 'Crinoidea,' and designated by Müller, *Comatula Milleri.* I have reason to think that the latter—not the former, as believed by Edward Forbes—is identical with the *Comatula Mediterranea* of Lamarck. But this I cannot take upon myself certainly to say, until I have had a specimen of the *C. Mediterranea* for careful examination. Putting aside other differences, the distinction between the two species of Lamlash bay is well marked by the *presence* in *C. Milleri* of what Miller has described and figured as a single inter-radial plate in each of the angles between the origin of the rays, but which is really formed of *three* minute plates, which I suspect not to be inter-radials at all, but dermal plates; of these there is *no trace whatever* in the *Comatula rosacea.*

Table I.—Mollusca.

Species	Remarks
*Aclis nitidissima, . .	The whole of the shells named in this table were dredged in Lamlash Bay—the best ground being the area that extends from Clachland Point and Hamilton Rock to the north end of Holy Island, in from 15 to 40 fathoms.
*—— unica, . . .	
Acmæa testudinalis, .	
——— virginea, . .	
*Adeorbis sub-carinata, .	
*Amphysphyra hyalina, .	
Anomia ephippium, .	
——— aculeata, . .	
Aplysia hybrida, . . .	In the pools, Clachland Point.
Aporrhais pes-pelicani, .	
*Arca lactea, . .	
Artemis exoleta, . .	
———- lincta, . .	
Astarte compressa, .	
*——— elliptica, . .	
*——— Scotica, . .	
——— sulcata, . .	
Buccinum undatum, .	
*Cœcum glabrum, . .	
*——— trachea, .	

Species	Remarks
Cardium echinatum, .	
——— edule, . .	
*——— fasciatum, .	
*——— nodosum, .	
——— Norvegicum, .	
*——— Suecicum, .	
*Cerithiopsis tubercularis.	
*Cerithium adversum, .	
——— reticulatum, .	
Chiton asellus, . .	
——— marmoreus, .	
——— ruber, . .	
*Chemnitzia indistincta, .	Scarce in Lamlash Bay.
*——— rufescens, .	
Circe minima, . .	
Corbula nucleus, . .	
Crania anomala, . .	Attached to stones and shells in deep water—not scarce in Lamlash Bay.
*Crenella decussata, .	
*——— discors, . .	
*——— marmorata, .	
Cylichna cylindracea, .	
*——— mammillata, .	
*——— obtusa, . .	
*——— truncata, .	
*——— umbilicata, .	
Cypræa Europæa, .	
Cyprina Islandica, .	Dead shells, not uncommon.
Dentalium entalis, .	
*Donax anatinus, . .	
Emarginula reticulata, .	
Eulima bilineata, . .	Good specimens from gravel bottom, at about 15 fathoms, north end of Holy Island.
——— distorta, . .	
* ——— polita, . .	
Fissurella reticulata, .	
Fusus antiquus, . .	
——— Islandicus, .	
Kellia rubra, . .	Concealed in the plant *Lichina pigmæa*, near the shore, Holy Island.
——— suborbicularis, .	In different parts of Lamlash Bay.
*Lacuna pallidula, . .	
*——— vincta, . .	
Lamellaria perspicua, .	
*Leda caudata, . .	
*Lepton convexum, .	

Lima hians, . . . —— loscombii, . . —— subauriculata, .	At the north end of Holy Island, in about 10 or 15 fathoms. The nests of *Lima hians*, with the live shells, are very abundant in this locality; and they are worth examining for *Diatomaceæ*.
Littorina littoralis, .	
——— littorea, . .	
——— neritoides, .	
——— rudis, . .	
*Lucina flexuosa, . .	
*—— spinifera, . .	
*Lucinopsis undata, .	
*Lutraria elliptica, . .	
Lyonsia Norvegica, .	
Mactra elliptica, . . —— solida, . . —— stultorum, . . —— subtruncata, .	In the sands, low water, Lamlash.
Mangelia costata, . . ——— Leufroyi, . *——— linearis, . . *——— nebula, . .	In deep water between Clachland Point and Holy Island.
———rufa, *var. Ulideana* *——— septangularis, . ——— teres, . . *——— turricula, .	*M. Rufa* was dredged between Fullarton's Rock and King's Cross Point.
Modiola modiolus, . .	
*—— phaseolina, .	
*Montacuta bidentata, .	Attached to the spines of *Spatangus purpureus*.
——— substriata, .	
*Mya arenaria, . .	
*—— truncata, . .	
Mytilus edulis, . .	Immature specimens everywhere.
Nassa incrassata, . .	
—— reticulata, . .	
*Natica alderi, . .	
—— monilifera, . .	
—— Montagui, . .	
*Nucula nitida, . .	
*—— nuaeus, . .	
*—— radiata, . .	
*Odostomia conoidea, .	
*——— cylindrica, .	
*——— decussata, .	
*——— excavata, .	
*——— interstincta, .	

Species	Locality
*Odostomia plicata, .	
*———— rissoides, .	
*———— spiralis, .	
*———— unidentata, .	
Ostrea edulis, . .	An oyster-bed, Lamlash Bay, Holy Island.
*Ovulum acuminatum, .	
Patella athletica, . .	Holy Island.
——— pellucida, . .	
——— vulgata, . .	
Pecten maximus, . .	Lamlash Bay.
——— opercularis, .	Lamlash Bay.
*——— pusio, . .	Lamlash Bay.
——— similis, . .	Lamlash Bay.
——— striatus, . .	Lamlash Bay.
——— tigrinus, . .	Lamlash Bay.
*——— varius, . .	Lamlash Bay.
Pectunculus glycimeris, .	
Philine aperta, . .	
*——— catena, . .	
*——— scabra, . .	
Pholas crispata, . .	In the sandstones, low water, Lamlash.
Pileopsis Hungaricus, .	
Pilidium fulvum, . .	
Pinna pectinata, . .	
Pleurobranchus membranaceus (?), . .	Under stones, low water, Holy Island.
*Pleurotoma gracilis, .	
*———— septangularis,	
Psammobia Ferroensis, .	
Puncturella Noachina, .	
Purpura lapillus, . .	
*Rissoa Beanii, . .	
*——— calathus, . .	
*——— cingillus, . .	
*——— costata, . .	
*——— costulata, . .	
*——— crenulata, . .	
*——— fulgida, . .	
*——— inconspicua, .	
*——— interrupta, . .	
*——— labiosa, . .	
*——— parva, . .	
*——— punctura, . .	

* *Pinna pectinata* was also dredged by the late Major Martin on Skelmorlie Bank, near Wemyss Bay, Ayrshire coast.

*Rissoa rubra, . .	
*—— rufilabris, . .	
*—— semistriata, .	
—— striata, . .	
—— striatula, . .	
*—— ulvæ, . . .	
*—— vitrea, . .	
*—— Zetlandica, .	
*Saxicava arctica, . .	
*—— rugosa, . .	
*Scalaria communis, .	
Scaphander lignarius, .	
*Skenea divisa, . .	
*—— nitidissima, .	
*—— planorbis, . .	
*—— rota, . . .	
*Sphœnia Binghami, .	
*Syndosmya prismatica, .	
Tapes aurea, . . .	
—— decussata, . .	
—— virginea, . .	
*Tellina crassa, . .	
—— donacina, . .	
*—— fabula, . .	
*—— incarnata, . .	
*—— solidula, . .	
*—— tenuis, . .	
Terebratula caput-serpentis	Deep water, between Hamilton Rock and Holy Island.
*Teredo megotara, . .	
—— Norvegica, .	In the wreck of the pier, Holy Island,
Thracia phaseolina, .	
—— pubescens, .	
Trichotropis borealis, .	
Trochus cinerarius, .	Lamlash Bay. *T. umbilicatus* is common along the shore of Holy Island, also *T. cinerarius.*
—— magus, .	
—— millegranus, .	
—— Montagui, .	
—— pusillus, . .	
—— tumidus, . .	
—— umbilicatus, .	
—— undulatus, .	
—— ziziphinus, .	
*Trophon Barvicensis, .	
*—— clathratus, .	
Turritella communis, .	
*Turtonia minuta, . .	

Species	Remarks
Venus casina, . .	North end of Holy Island, in from 10 to 20 fathoms.
—— fasciata, . .	
—— gallina, . .	
—— ovata, . .	
—— striatula, . .	

TABLE II.—NUDIBRANCHIATE MOLLUSCA.

Species	Remarks
Doris bilamellata, . .	Under stones at low water, in many places, and not uncommon. *D. planata* was found in Lamlash Bay by Mr Alder.
—— planata, . .	
—— tuberculata, . .	
*Eolis alba, . . .	
*—— coronata, . .	
—— Drummondi, .	Under stones at low water, in many places, and not uncommon.
*—— Landsburgii, .	
Goniodoris nodosa, .	
Lomanotus —— (?), .	Probably a new species, dredged in Brodick Bay by Dr Greville and Dr Miles, in about 15 fathoms, between Invercloy and Corriegills, but the specimen was unfortunately lost before it could be sent to Mr Alder for examination. It was two inches in length, white, with orange processes, and of rare beauty. A sketch taken by Dr Greville was forwarded to Mr Alder. *L. flavidus* was dredged in 1846 by Mr Alder in Lamlash Bay.
———— flavidus, .	
Polycera quadrilineata, .	
Triopa claviger, . .	

TABLE III.—CRUSTACEA.

Species	Remarks
Balanus balanoides, .	The *Cirripedia* have scarcely been examined.
Cancer Pagurus . .	The edible crab is common around Arran.
Carcinus Mœnas, . .	
*Ebalia Bryerii, . .	Dredged by Dr Greville in Lamlash Bay.
*—— Cranchii, . .	
—— Pennantii, . .	Not uncommon in deep water.
Eurynome aspera, . .	Several specimens obtained from deep water, Lamlash Bay.
Galathea squamifera (?),	

Gonoplax angulata, .	On the Ayrshire coast, abreast of Arran.
Hippolyte varians, .	
Homarus vulgaris, . .	The common lobster is found among the rocks on different parts of the Arran coast and at Holy Island.
Hyas araneus, . .	
* —— coarctatus, . .	
Inachus Dorsettensis, .	
*Lepas anatifera, . .	Found in Lamlash Bay on floating wreck.
Lithodes Maia, . .	In deep water, mid-channel, off Arran.
Pagurus Bernhardus, .	Abundant everywhere.
——— Prideauxii, .	Abundant in Lamlash Bay, with *Adamsia palliata.*
Palæmon squilla, . .	Common in all the rock-pools.
Palinurus vulgaris, .	Obtained by the late Major Martin from off Campbelton.
Pandalus annulicornis, .	
*Pinnotheres pisum, .	
Porcellana longicornis, .	Among the roots of *Laminaria.*
——— platycheles, .	Abundant under stones at low water, Holy Island and Lamlash.
Stenorhynchus phalangium,	Dredged in deep water.

Table. IV.—Echinodermata.

Amphidotus cordatus, .	The following species were all obtained in Lamlash Bay.
*——— roseus, .	
Asterias aurantiaca, .	Near Fullarton's Rock.
Asterina gibbosa, . .	In pools, north end of Holy Island, etc.
Chirodota digitata, .	Dredged near the pier, Holy Island.
Comatula rosacea, . .	Abundant in Lamlash Bay, near Holy Island, in from 8 to 15 fathoms. Solitary individuals have been traced up the Clyde as high as Hunter's Quay, Dunoon.
——— Milleri, . .	In Lamlash Bay, rare.
Cribella oculata, . .	
——— rosea, . .	This beautiful star-fish (two specimens) was dredged in deep water, north end of Holy Island.
Echynocyamus pusillus,	
Echinus miliaris, . .	
——— sphœra, . .	
Goniaster Templetoni, .	Not uncommon in the Bay.
Luidia fragilissima, .	Not uncommon in the Bay.

Ophiocoma bellis, . .	
———— granulata, .	
———— rosula, . .	
Ophiura texturata, .	
Palmipes membranaceus,	Good specimens, deep water, near Hamilton Rock.
Sipunculus ———— (?),	
Spatangus purpureus, .	
Solaster papposa, . .	
Uraster glacialis, . .	
——— rubens, . .	

TABLE V.—ZOOPHYTA.

Actinia bellis, . .	In the rock-pools, not uncommon.
——— coriacea, . .	Under ledges of rocks, in pools, etc.
——— crassicornis, .	Under ledges of rocks, in pools, etc.
——— mesembryanthemum,	
Adamsia palliata, . .	Abundant in Lamlash Bay, attached to shells occupied by *Pagurus Prideauxii.*
Anthea cereus, . .	
Antennularia antennina,	
*———— ramosa, .	
Cellepora pumicosa, .	
———— ramulosa, .	
———— Skenei, . .	
Campanularia dumosa, .	
Flustra foliacea, . .	
Halecium halecinum, .	
*Halichondria panicea, .	
*———— suberea, .	
Laomedea geniculata, .	
Lepralia annulata, .	On stones and dead shells, in deep water, Lamlash Bay.
——— hyalina, . .	On stones and dead shells, in deep water, Lamlash Bay.
——— Malusii, . .	On stones and dead shells, in deep water, Lamlash Bay.
——— Peachii, . .	On stones and dead shells, in deep water, Lamlash Bay.
——— trispinosa, .	On stones and dead shells, in deep water, Lamlash Bay.
——— violacea, *var.* cruenta,	On stones and dead shells, in deep water, Lamlash Bay.
Plumularia pinnata, .	
Salicornaria farciminoides,	
Sertularia tamarisca,	

ENTOMOLOGY OF ARRAN

LEPIDOPTERA.

115. AMONG the Lepidoptera, or tribe of butterflies and moths, occurring in the Isle of Arran, are several of rather peculiar interest; and a work treating of the Fauna of Arran would be incomplete without some notice of them.

COLIAS EDUSA (the clouded yellow butterfly) is a species which is very seldom seen so far north; but its occurrence near Lamlash was chronicled in the *Zoologist* for 1848, p. 1985.

EREBIA BLANDINA (the Scotch Argus) is a mountain species of butterfly, which, though common in many northern localities, is esteemed a prize by all collectors of the plains, when first they meet with it. It is common on many of the hill-sides in Arran.

EREBIA LIGEA is a species closely allied to BLANDINA, which has been reported to occur in the Isle of Arran; and Mr Curtis, the distinguished author of *British Entomology*, assured me that the late Sir Patrick Walker told him he had himself taken it in the island, in the second half of August. No recent captures of this insect in Britain are known; and the fact of its having ever been caught here being much doubted, the species has ceased to figure in our lists of indigenous species.

CŒNONYMPHA DAVIES is common on boggy places at some altitude above the sea.

Of the handsome genus of FRITILLARIES three species—ARGYNNIS AGLAIA, ADIPPE, and SELENE—are by no means uncommon in the little glens, up the hill-sides, and in early spring THECLA RUBI (the green hair streak) is frequent amongst bramble bushes.

POLYOMMATUS ARTAXERXES is a species very likely to occur in Arran, though I am not aware that it ever has been found there.

ANTHROCERA MINOS.—The recent capture near Oban of this species of *Sphinx*, found three years ago in Ireland, renders it certainly not improbable that it may occur in Arran. Like many of its congeners, it is excessively local, and may occur in some limited spot, only on one hill-side, yet there be in profusion.

Among the family of the day-flying moths, the BOMBYCINA, none of those which have been observed in the island are of sufficient importance to deserve special notice.

Amongst the NOCTUINA, the pretty THYATIRA BATES (peach-blossom) is common; CELÆNA HAWORTHII is tolerably plentiful; HADENA ASSIMILIS has once or twice been taken on the face of stone rocks; and PLUSIA INTERROGATIONIS is rather common than otherwise.

Amongst the GEOMETRINA which have been noticed here, may be mentioned MÆSIA BELGIARIA, which flies in heathy places; OPORABIA FILIGRAMENARIA, LARENTIA CÆSIATA, SALICARIA, and the mountain species of the prettle little genus EMMELESIA, viz.: TÆNIATA, ERICETATA, and BLANDIATA, and the pretty insect CARSIA IMBUTATA, of which the Scotch specimens are so much more delicately marked than those from the neighbourhood of Manchester.

Of the smaller species of Lepidoptera none of peculiar interest have yet been noticed in Arran; but it must be borne in mind that this portion of its natural treasures has never been thoroughly explored.

LIST OF LEPIDOPTERA COLLECTED IN 1836 BY DR CONNELL:*—

Cynthia cardui.
Hipparchia blandina.
——— polydama.
——— pamphilus.
——— hyreranthus.
——— janira.
——— semele.
Polyommatus alsus.
——— Alexis.
Vanessa urticæ.
Pontia brassicæ.
——— rapæ.
——— napi.
Melitiæa Euphrosyne.
Argynnis Aglaia.
Lycæna phlœas.
Arctia caja.
Cerura vinula.
Macroglossa stellatarum.
Minoa chœrophyllata.
Anthrocera filependulæ.
Spilosoma menthastri.
Hypena proboscidalis.
Mamestra brassicæ.
Hepialus velleda.
——— humuli.
Rumia crotsegata.
Ophiusa lusoria.
Leucania pallens.
Plusia chrysitis.
——— gamma.
Episema cœruleocephala.
Leucania impura.
Larentia chenopodiata.
Fidonia atomaria.
Cabera exanthemata.
——— pusaria.
Actebia porphyrea.
Xylina putris.
Harpalyce fulvata.
——— sylvatica.
——— ocellata.
Margaritia verticalis.
Botys forficalis.
Anarta myrtilli.
Pterophorus punctidactylus.
Nemeophila plantaginis.

* This list contains all that has been published up till this time regarding the Entomology of Arran. It was furnished to Dr Landsborough by the late Dr Connell of the High School of Glasgow; and was published first in the *New Statistical Account of Scotland*, Vol. V., and afterwards in Dr Landsborough's *Excursions to Arran*.

GEOLOGY OF CLYDESDALE

116. Few parts of Scotland present so rich a field of geological inquiry as the district of Clydesdale. In none certainly are the facilities so great for studying the varied phenomena of a large class of formations. Even within the compass of its capital city, monuments exist of many successive revolutions in physical geography and organic life; beneath her streets are entombed the remains of a vast extinct creation; the foundations of her teeming warehouses and crowded quays are laid amid the mouldering works of a Primeval Race; while between these extremes there is interposed a long series of terms indicating a progressive advance towards the existing forms of animal life and aspects of the surface. Of scarcely less interest are the shores of the noble Frith, which present a complete suit of the older formations; while among its islands Arran has long been celebrated as exhibiting a greater variety of geological phenomena than perhaps any other tract of like area on the surface of the globe.

In the following pages a few brief notices are given of the various deposits, their fossil contents, and the best points for studying the relations of the strata, in order to direct the researches of future inquirers, and to aid those who are beginning the study of the science.

I.—HUMAN PERIOD.

117. The city of Glasgow stands on the great coal field of Scotland, about seven miles from its north-west border. Within its bounds the several "measures" of this formation,

sandstone, shale, clay, ironstone and coal, crop out in many places; but the surface is generally covered with newer deposits. The site of the smaller portion on the south side of the river is a level plain, composed of ancient estuary or fluviatile accumulations. A narrow strip on the north, throughout the whole length of the city, is of the same character. North of this, the ground rises with considerable rapidity in a series of elliptic-shaped hills, from 100 to 250 feet above the plain, their longer axes being parallel to the course of the river, here west-north-west. The principal heights are Garngad hill, 252 feet 6 inches; Necropolis, 225 feet; Garnethill, 176 feet 7 inches; Blythswood hill, 135 feet 3 inches; Woodlands hill, 153 feet; Hillhead, 157 feet; Observatory, 179 feet 5 inches;* Jordanhill, two miles west, 145 feet. Some of these were originally a little higher, the tops having been levelled to afford a broader space for the erection of buildings. The elevations give a pleasing variety to the city, and afford striking views of the distant Highland mountains, and the fine amphitheatre of nearer hills, which close in, on all sides except the east, the undulating and well-wooded basin intersected by the lower course of the Clyde.

118. Apart from deposits now in progress, the latest formation is that of the estuary and fluviatile accumulations above referred to. These are spread out continuously on both sides of the river, but more widely on the south side, from Rutherglen, three miles above Glasgow, to near Erskine, ten miles below it—the former being the upward limit of the tide before it was obstructed by works connected with the improvement of the navigation—the latter the termination of the river in the estuary, and the limit of the ascent of salt water. The deposit consists throughout of laminated beds of sand and loam, with thin courses and streams of gravel and layers of peat. Marine shells have been found sparingly in its middle and lower parts, and a few fresh water species in the upper beds; but no collection has been formed of these

* These heights are given on the authority of Mr Thomas Kyle, Civil Engineer.

interesting objects, so that we cannot describe the fossil contents more minutely. The deposit has been tranquilly formed throughout, long periods of repose having been but rarely interrupted by floods. Ancient rude canoes have been found in various parts of it, deeply imbedded in the sand and loam, one at either end of the area, and a great many on the banks of the river at Glasgow; some at heights 10 or 12 feet above the highest level reached by the greatest floods on record in the Clyde. Respecting these, the following particulars have

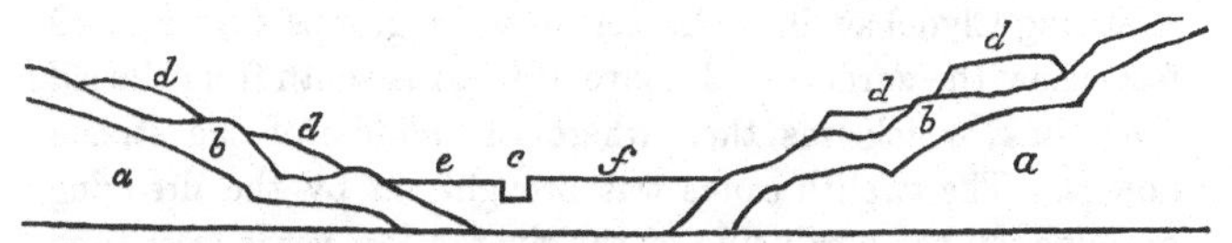

(a a) Carboniferous beds; (b b) Till and lower sand beds; (c) river Clyde; (d d) post-pleiocene, or raised beaches; (e f) estuary deposits of the human period.

been kindly furnished by John Buchanan, Esq., the well-known archæologist of this city, who carefully noted at the time the circumstances attending the discovery of those more recently found:—

"Within the last eighty years, no less than sixteen of these interesting remains of aboriginal workmanship have been found in and near Glasgow. They are all, with one exception, formed of single oak trees, in some instances by the action of fire, in others by tools evidently blunt, probably of stone, and therefore referable to a period so remote as to have preceded the knowledge of the use of iron. The first known instance was in 1780. The canoe lay under the foundations of the old St Enoch's Church, at a depth of 25 feet from the surface—that is, about the level of low water in the river below Argyle Street—and within it was a stone hatchet of polished greenstone, in good preservation. It is now in the possession of James Smith, Esq., of Jordanhill. The second, in 1781, while excavating the foundations of the Tontine, at the Cross, the surface being here 22 feet above high water. A third, in 1824, in Stockwell Street, in a deep cutting opposite the mouth of Jackson Street.

The fourth was found, in 1825, in the cuttings for a sewer in London Street, on the site of the 'Old Trades' Land:' the canoe was vertical, the prow uppermost, and a number of shells were inside. The next discovery was made in 1846, when the improvements in the river began to be actively carried out. Eleven canoes were discovered in a short period. Of these, five were found in the lands of Springfield, opposite the lower portion of the harbour; five more on the property of Clydehaugh, west of Springfield; and one in the grounds of Bankton, adjoining Clydehaugh. The ten were in groups together, 19 feet below the surface, and above 100 yards south from the *old* river bank, which was then where the middle of the stream now is. The twelfth canoe was brought up by the dredging machine, on the north side of the river, a few yards west from the Point House, where the Kelvin enters. The Erskine specimen was found in 1854. It was taken out by Mr Taylor, who has charge of the ferry, nearly entire. To test its capabilities, he had it partially supported on a raft, and floated in it across the stream.

"A collection of these canoes is now preserved in a building in the College grounds; and single specimens may be seen in Stirling's Library, Miller Street; the Andersonian Museum, George Street; Ferry-house, Erskine ferry, ten miles below Glasgow; and in the hall of the Society of Antiquaries, Edinburgh." *

119. The conclusion is forced upon us by these facts that the entire area was at a remote time covered by an estuary, connected with the sea by a narrow strait near Erskine, where the hills on either side press close upon the stream; whose limits reached inland almost as far as Johnstone and Paisley, narrowed upward by the projecting Ibrox and Pollokshields ridges, but again widening out, so as to wash the base of the Cathkin and Cathcart Hills, and sweeping round north-east in a wide bay, so as to cover the space now occupied by the

* Mr Buchanan has since published a very full and interesting account of these curious remains in *Glasgow Past and Present*, Vol. iii.

Glasgow Green and suburbs of Bridgeton. The river then entered about Uddingston or Rutherglen; and the northern shore was formed by the lower slopes of the hills already alluded to, and their continuations north-west by Partick, Jordanhill, and Yoker, to the vicinity of Erskine. The period to which the canoes are referable, and when Clydesdale had the features just sketched—probably the "stone period" of Scottish archæologists—lies far back in pre-historic time; but how far we have no means of knowing. We know, however, that nearly 2,000 years ago, the Roman wall was constructed between the Forth and Clyde—from Bowling to Grangemouth; and, as Mr Smith of Jordanhill has happily pointed out, no oscillation in level has taken place since that time. This singular work had precise reference even to the present tide levels.* How remote, then, must be the time when the quiet waters of the estuary laved the hill-sides, now covered by busy thoroughfares; and a race, whose other memorials are lost, navigated in these rude canoes the broader waters of the river, whose narrowed stream now floats the largest ships, and brings to our doors the choicest products of the globe.

II.—PLEISTOCENE OR GLACIAL PERIOD.

120. The beds just described overlie towards their margin another series, occupying a peculiar place, and presenting a marked organic sequence, which links on the tertiary age to the existing order of things, and affords another, amid the now oft-recurring examples of passages from group to group, which almost defy classification, and shew us how past creations shade off into the present, continuously, without a chasm. The deposits in question are well known to geologists as the Clyde Beds, the discovery of which, with an interesting account of their natural history, we owe, as before stated (Art. 21), to Mr Smith of Jordanhill. This series of beds, while very

* In opposition to the view here stated, Mr Geikie has attempted to shew in a late paper (*Jour. Geol. Soc.*, Vol. xviii., p. 218) that there has been an oscillation of level in the recent period. But we cannot regard as conclusive the evidence which he has adduced.

similar to the Arran series already described (Arts. 20–22, 86–91), has a more complete development. In the basin of the Clyde there is interposed between the boulder-clay and the Arctic shell bed a bed of fine laminated clay not found in Arran; and in the position of the upper drifts of Arran there occurs a shell bed, with existing British species, which have been as yet found in Arran only at the base of the old sea-cliff (Arts. 22, 69). The Clyde series, in its normal state of development, consists of the following beds in descending order:—

1. Gravel, sand, or clay, and alluvial soils.
2. Sand and gravel, with British species of shells.
3. Clay without shells.
4. Clay with Arctic species of shells.
5. Fine laminated clay, without shells.
6. Boulder-clay at the base, over the natural rock.

There are thus two shell beds—one of Arctic, another of British species—separated by a bed of clay; each is characterised by a particular species, whose occurrence is so constant that the lower bed, No. 4, may be called the *tellina proxima* bed, and the upper, No. 2, the *cardium edule* bed. It has been already stated that no stratified beds occur in Arran below the boulder-clay, which rests directly on the rock; and the same arrangement exists throughout the entire basin of the Clyde, with the single exception of a case mentioned by Mr Smith of Jordanhill, to which allusion will again be made. There are, however, two or three cases besides, recorded by competent observers, in other parts of Scotland.

These beds will now be described in a descending order. As often stated already, the last event in the physical history of Arran was the carving out of the old sea-cliff, during the progress of an elevation of the land, amounting to 25 feet, and the same remark applies to the whole region of the Clyde. With this and with still higher terraces shell beds are connected, marking probably the sea levels of an earlier period, and attesting by their wide distribution that general elevation of the land, the beginning of which seems to have been coincident with the disappearance of the ice of the glacial epoch. It is in this view of their origin that these upper shell beds are often termed "Raised beaches."

121. When we pass from the river Clyde into the frith at Dunbarton, both shores are seen to be marked by a well defined terrace, 10 to 20 feet above the present tide level, and bounded inland by a steep cliff. The Greenock Railway, from the Bishopton tunnel to Greenock, runs along this terrace; and it is equally well defined upon the Cardross side, from the Leven mouth by Helensburgh to Gareloch-head. The watering places of Gourock, Kempoch, and Ashton, and on the opposite side those of Kilcreggan, Cove, and Peyton, stand upon it. In Roseneath peninsula it is extremely well defined and traceable round the whole shore. Everywhere, indeed, upon the shores of the frith and its islands the same terrace is clearly marked; nor less so on the west coast, throughout Cantire, from the Crinan canal, by Oban, up the shores of Loch Linnhe, on Loch Fyne, etc., and along it on most parts of these shores the coast road is carried. The cliff, which bounds it inland at a distance varying from a few yards to a quarter of a mile, and even half a mile, as in parts of the Roseneath and Renfrewshire shores, is everywhere sea-worn, and hollowed out into caves; the terrace is covered with shingle greatly sea-worn; it is flat and difficult to drain; the elevation seldom passes 40 feet; usually it is considerably less. Mr Smith was the first to point out that at the level which the land then had it must have stood for a much longer period than that which has elapsed (2,000 years) since the Roman works in this country were constructed. The great length of the caves, of which Professor James Nichol has given a remarkable instance on Davar island at Campbelton*—as well as other cases of wearing, such as the projection of the harder veins in the rocks, the overhanging of the cliffs, etc., clearly prove this. In these caves and on the terrace, sea-shells of species now existing in

* *Jour. Geol. Soc.*, Vol. viii., 1852.

the adjoining sea, occur abundantly and in many places. One of the most remarkable, as regards elevation, which has come under our notice is the steep terrace at Roseneath House, a seat of the Duke of Argyle, which is shewn in the annexed cut. The heights were kindly furnished by the late Mr Lorne Campbell, of Roseneath.

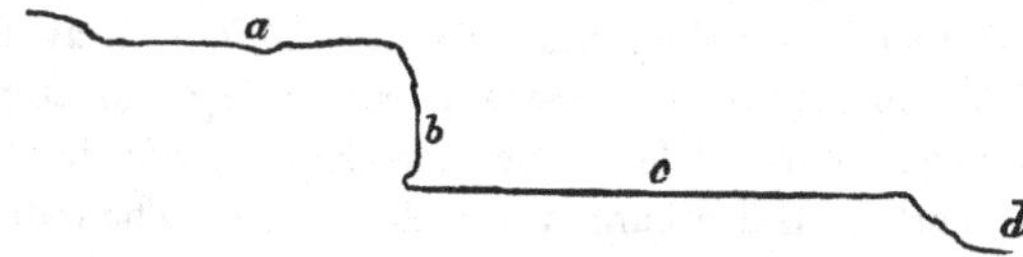

(a) Upper slopes, on which the offices stand; (b) Sea-worn cliff of old red sandstone, called Wallace's loup; (c) Terrace of former beach, on which Roseneath house stands; (d) Sea level.

The upper terrace is 79 feet high, the lower north portico of Roseneath House 42 feet; shells, broken and mixed with sea-weed, are found on both terraces, two or three feet below the surface; on the upper terrace in hollows 68 feet above high-water. At Johnstone, near Paisley, a case is mentioned by Mr Smith, in which sea-shells, bones of fishes and sea-birds, claws of crabs, and sea-weed were found at about 80 feet elevation, resting on till beds 70 feet thick. The brickfields about Glasgow and Paisley and other parts of the Clyde shores abound in these shells. The Paisley fields are to the north-west of the railway station; they yield a good many species, all of a littoral character. Similar beds are found on many parts of the coast, and in some other cases besides those already mentioned, at considerable distances inland. Such inland localities, however, are always so situated as to have had, while former high levels of the water prevailed, open communication with the frith. The late Dr Landsborough found 47 species 15 feet above the sea level, at Stevenston, near Saltcoats; 68 in a bed south of Largs; and many species about Ardrossan. Such beds occur on the shores and islands of Loch-Lomond in several places, on the shores of Bute, and those of the outer friths; their distribution being no doubt co-extensive with the elevated margin of the frith of Clyde and its numerous branches. These beds have yielded in all about 200 species; the shells being generally of a littoral character.

The higher grounds of the interior of Clydesdale, up to the height of more than 1,000 feet, are occupied in many places by stratified drifts, newer than the boulder-clay, and consisting of water-worn materials, varying from a coarse gravel and sand, with large stones, to fine sediment of sand and clay; but wholly unfossiliferous. These materials are often arranged into long ridges, broad-based, but narrow at top, and 50 to 70 feet above the adjoining ground. They are locally termed Kames, and are similar to the Oäsars of Sweden, the Eskers of the central tracts of Ireland, and the Drums of the north. They are more common in open districts, and affect a particular direction—generally from north or north-west to south or south-east; and in our experience usually shew rolled fragments, whose origin can only be looked for towards the north-west. The circumstances under which these upper drifts were probably formed have been already indicated (Art. 90). Ice and water have both combined their forces in the production and final arrangement of this drift—ice most probably in its first transport; currents when the land was rising, re-adjusting the materials and giving the existing forms. Many of the stones still preserve their striations, and hence the action of such currents could not have been very long continued.

122. The beds which follow these, as we descend, possess a much greater degree of interest. They are—

1. Clay, varying in many places to sand and gravel.
2. Clay, with Arctic shells.
3. Fine laminated clay.
4. Boulder-clay, or till.

The uppermost of these beds varies greatly in different localities; the three beds below it are more constant in character and persistent in extent. The bed, No. 1, is worked as a brick and tile clay in many parts of Clydesdale; but the underlying clay is rendered unworkable by the shells which it contains, especially the large valves of *Cyprina Islandica.* The clay of this bed is sometimes sandy, sometimes compact and tough. In some places it is free of stones, in others these are thickly scattered through it; and in rare cases the stones are found striated. The Arran shell bed is more stony; and upper stony drifts represent there the upper Clyde clays. But perhaps the most remarkable feature of the Clyde series, as contrasted with that of Arran, is the constant presence of the bed No. 3, the fine laminated clay, which divides the shell-bed from the boulder-clay. It is a fine fissile clay, easily opening into thin *laminæ*, like the leaves of a book; it is unfossiliferous and entirely without stones. It is thus remarkably contrasted with the boulder-clay on which it rests. We have seen that it is absent in Arran—the shell bed resting on the boulder-clay directly. The latter has very much the same structure and composition in every part of the country which it exhibits in Arran; and these have already been sufficienty described. It is widely distributed over Clydesdale. With a preponderance of local stones, it has everywhere many that are from a remote origin, and these are distributed throughout the mass, without reference to their weight, to a near origin, or distant transport. Many cases, however, occur in which the larger stones are below and the smaller in the upper layers. The boulder-clay forms the covering of all the hills and slopes in and around Glasgow above the levels of the ancient estuary (Art. 119), and the foundations of the houses of all the higher streets within the city are laid in it. So varied are the contents of these till beds within the limits of Glasgow that a tolerably complete set of Scottish rocks can be made from them; granite, gneiss, mica schist, clay and chlorite slates, quartz rock, porphyries, old red sandstone and conglomerate, limestone, coal, sandstone and shale, ironstone, and a great variety of traps, are all turned out from the same cutting. The excavations on Woodlands hill for the roads and buildings of the new park exposed such assemblages. A large collection was made by a geologist of Glasgow from the foundations of the houses of Windsor Terrace West, and exhibited at the last meeting of the British Association in that city. A noted locality which excited much interest during the meeting of the Association in 1840, and has been often since referred to, was Bell's Park, north-east side of the city, near the Caledonian Railway Station. From such sections as these striated, grooved, and polished boulders will be obtained even more readily than in the terminal moraine of a Swiss glacier.

In Clydesdale, as in Arran, the thickness of the boulder-clay varies greatly—sometimes but a few inches, in other cases reaching 40 and 50, and even 60 or 70 feet; as a rule, it is thickest in the valleys and thins out over the tops of the ridges (Art. 21); it is generally thinner and more frequently absent from trap hills than from those composed of sedimentary formations, probably from the latter yielding easily and abundantly the clay which forms the base. In rare cases, at heights reaching close on 1,000 feet, the thickness is from 40 to 60 feet. It is also known to extend in many cases below the level of the Atlantic and German oceans. It has been stated already that the boulder-clay is unstratified, but a few cases are on record of stratified materials being found in the deposit. Some of these are probably cases of mere apparent interstratification. The upper surface of the boulder-clay undulates very much; and in the hollows between the undulations, the next bed, the laminated clay, comes on, and so appears lower than the boulder-clay at either side. If the upper part of such a section be concealed, this arrangement will appear to be a case of true interstratification. Another source of error is the cause already sufficiently illustrated in Art. 85. There are, however, one or two cases of this interstratification recorded by Mr Geikie, from his own observation,* and we can hardly suppose that an observer so experienced should be mistaken. These cases occurred at

* "The Glacial Drift of Scotland." Glasgow: Gray. 1863.

Crowbyres bridge, Slitrig water, a quarter of a mile from Hawick; at Garpal, in Ayrshire, where the Whitehaugh water joins the Ayr; close to the parish church of Carmichael, Lanarkshire; and at Chapelhall, near Airdrie.* Stratified beds of sand and clay, or sand and gravel, are said to occur in a very few cases below the boulder-clay, between it and the natural rock. We have already remarked (Art. 90) that no such case has come under our notice; but we do not doubt the competence of other observers.

123. Having thus traced the physical relations of these beds, in order to preserve continuity in the descriptions, we have now to consider their fossil contents. The shell bed No. 3 is the main if not sole repository of these fossils. This bed has yielded a great number of species, and of these from 10 to 16 per cent. are either extinct or not now known in the British seas. Those which are recent, but not found in the British seas, occur in the Arctic regions, and indicate, therefore, the existence of a colder climate than that of the present period, and colder also than that under which the shell beds of the upper drifts, or "raised beaches," were accumulated. In remarkable contrast to the state of the shells in these upper drifts, the shells of the lower or Arctic shell bed are generally entire in the beds throughout Clydesdale, even delicate specimens being in good preservation. Such, however, is not the condition of the shells of the same Arctic bed in Arran. We have already seen (Art. 86) that they are in most cases in a far less perfect state of preservation.

The shelly deposit may be studied in many localities, accompanied in most cases, as already stated, by the two beds which underlie it. It is found at Paisley and Houston, and westwards along the shores of the river and frith, often between half-tide level and high-water mark. There are beds at Langbank, Greenock, and Gourock; on the opposite shore at Dalmuir, near Erskine; on Loch-Lomond shores and islands; the

* Mr Smith, of Jordanhill, long ago recorded a case, to which reference will again be made.

Gareloch, Holy Loch, and Loch-Long; in the Kyles of Bute, and Kilchattan bay; at Stevenston, near Saltcoats; and doubtless in many other places on both shores not yet discovered. There are traces of it on the western shore of Bute, and its extension into Arran has been already noticed. The elevation of the shell bed seldom reaches 50 feet; but on Loch-Lomond shores it is said to reach 70 feet; and the bed in the brickfields east of Jordanhill is about 80 feet. Balnacaillie bay, on the north shore of Bute, opposite the Burnt Isles in the Kyles, is one of the richest localities. It may also be called the classic locality, as it was from an examination of the shells found here that Mr Smith of Jordanhill was led to recognise the Arctic character of the deposit. A stream on the Bute side has here cut into the bed and exposed the shells; the bed extends a considerable way inland up a hollow, and along the level shore. In such a locality great care must be taken to distinguish the shells of the ancient bed from those of the present sea, which may have been introduced into its upper surface. The caution given in Art. 90 should also be attended to, especially if there be shell beds near of the newer age. There is less danger in a clear vertical section of the beds; but here the caution of Art. 85 will have to be carefully observed. In this locality Mr Smith found the following species:—

Cyprina Islandica.
Astarte Garensis.
——— elliptica.
——— multicostata.
Tellina proxima.
Pecten Islandicus.
——— opercularis.
Natica clausa.
——— glaucinoides.
Balanus costatus.
Serpula triquetra.
Spirorbis corrugatus.
Littorina littorea.
Lacuna vincta.
Modiola modiolus.
Fusus antiquus.
Nucula nucleus.
Leda oblonga.
Saxicava rugosa.
——— sulcata.
Panopæa Norvegica.
Mya udevallensis.
—— truncata.
—— arenaria.
Cardium edule.
Sphenia Swainsoni.*

* In this as in other lists, we have thought it better to give the names according to the original authorities. Conchologists will at once perceive that the nomenclature is not uniform.

To these Mr Crosskey has added the following:—
Natica Grœnlandica.
Mangelia pyramidalis.
——— var. Rosea (Sars).
Margarita undulata.
Crenella nigra.

This is the best locality for the *Panopœa Norvegica* and *Pecten Islandicus;* the latter occurs in large quantities, and perfect specimens are found with the Northern *Balanus* attached. One valve only of *Crenella* was found. *Mya udevallensis* occurs in the upright boring position in great numbers.

In a railway cutting between Port-Glasgow and Greenock, 50 feet above the sea level, Mr Smith found thirty-six species, of which half are no longer British. The section here, in descending order, was—

1. Alluvial soil.
2. Coarse gravel, 2 feet.
3. Sand, 10 feet.
4. Sand, gravel, and clay, full of shells.
5. Boulder-clay, depth unknown.

No notice is taken here of the laminated clay. It is by later observers that its persistency has been made out. The following is a list of the shells:*—

Cyprina Islandica.
Astarte Garensis.
——— multicostata.
Tellina proxima.
Pecten Islandicus.
Natica clausa.
——— glaucinoides.
Balanus costatus (?).
Littorina littorea.
Turbo expansus (?).
—— canalis (?).
Modiola modiolus.
Leda oblonga.
—— pygmæa.
Leda minuta.
Saxicava rugosa.
——— sulcata.
Trochus inflatus.
Mya truncata.
—— udevallensis.
—— ovalis.
Cardium edule.
Anomia ephippium.
Patella virginea.
Fissurella Noachina.
Fusus scalariformis.
—— Banffius.
—— discrepans.

* Smith's Tert. Geol., p. 32.

Fusus turricolus.
Buccinum undatum.
Mactra striata.
Lucina undata.
Amphidesma prismaticum.

In July of the present year Mr Smith and Dr Scouler have added *Bulla lignaria* and *Scalaria communis* to the list of species from Gourock.

At Dalmuir Dr Thomas Thomson * found twenty-nine species, afterwards increased to seventy by Mr Smith from a fresh excavation. At Stevenston Dr Landsborough found twenty-seven species in a bed of blue clay, 33 feet below the surface; of these, eight are no longer British.

Mr Smith estimates the entire number of species from the Clyde beds of this age at 151. The following list, kindly furnished by Mr Smith, contains all the species *not found in a recent state* in the British seas, being nearly 16 per cent. of the entire number found:—

Tellina proxima: Arctic seas.
Astarte multicostata.
——— Propinqua.
——— Withami: Bridlington: Arctic seas.
——— Borealis: Upper Crag: Arctic seas.
Mya udevallensis: Arctic seas.
Pecten Islandicus.
Leda oblonga.
—— antiqua.
Cytherea lævigata.
Mactra striata.
Natica clausa.
——— Smithii (Bulbus, Brown).
——— fragilis.
——— glaucinoides.
Nassa Monensis.
Buccinum striatum.
———— granulatum.
Trochus inflatus (Margarita undulata, Sowerby?): Arctic seas.

* Records of Gen. Science, i., 131.

Littorina expansa: Arctic seas.
Velutina undata: Arctic seas.
Fusus peruvianus: (Murex, Sowerby): Crag, Udevalla; Arctic seas.
Fusus imbricatus: Arctic seas.
—— curtus.
Saxicava sulcata, Udevalla.
Balanus udevallensis, Udevalla.

It is on the ground of the proportion of species extinct in our seas, as above stated, that Mr Smith, in conformity with the nomenclature of Lyell, called these beds Newer Pliocene or Pleistocene, that is, the newest tertiary; the beds resting on the upper drifts, with their existing species, coming, of course, under the post-tertiary or recent period. The series is clearly connected with a period of great cold. Active observers, following the course so ably marked out by Mr Smith, have been enabled to classify the superficial formations all over the north of Europe and in Canada, in conformity with the views which he has set forth.

124. Among those who have studied the Clydesdale beds since Mr Smith ceased to work actively upon them, the Rev. A. MacBride of Ardmory, Bute, and the Rev. H. W. Crosskey of Glasgow, have been the most zealous and successful. The Rev. W. Frazer of Paisley has done eminent service among the beds in his neighbourhood. Mr MacBride was the first to publish a division of the beds applicable to Bute, which has been since extended by various researches, and may now be considered as established for the Clyde beds. Mr Crosskey has lately found the same series of beds in the inland parts of the east of Scotland; and the division referred to, the same as we have already given, no doubt represents the series in its state of normal development in Scotland. There are, of course, variations, depending on local conditions, as there are among the members of the older formations; to these allusion has been already made. Mr Crosskey has, in the kindest manner, placed at my disposal all the facts recently ascertained by him in regard to

these beds, and has thus enabled me to give a record of their history complete to the present time.

Of the shells in Mr Smith's list, above given, the following are most abundant:—*Tellina proxima*, *Pecten Islandicus*, *Mya udevallensis*, *Leda oblonga*, *Natica clausa*, *Trophon Scalariformis*, *Saxicava sulcata* (large Arctic form), *Balanus udevallensis*. It is the profusion of these species, the existence of an abundant Arctic fauna *in situ*, which gives force to the argument for a former severe climate. There is another fact which places the character of these clays beyond doubt—the occurrence in them, in abundance, of northern forms which are rare in the waters of existing seas. There is, for example, the *Panopæa Norvegica*, the prize of the deposit for collectors, which has never yet been found living in the Clyde, but is abundant in the clays; it is taken very rarely on the Dogger bank, and only one living specimen has been found on the coast of Norway. The *Puncturella Noachina*, found sparingly in the Clyde, has its chief habitat in the northern seas, and is abundant in some of our clay beds. The same may be said of *Trichotropis borealis*, *Trophon clathratus*, and *Leda pygmæa*. That the deposit is in no sense a drift, but the repository of a fauna abundant on the spot, is manifest from many facts: the shells are in an exquisite state of preservation; the epidermis is often as well preserved as in living specimens; the most fragile bivalves are found by dozens, with both valves united; the shells are pretty uniformly distributed through the clays for miles in length; such borers as *Mya udevallensis* and *Panopæa Norvegica* are in their natural boring positions; and *Mya truncata* is found with the syphon preserved.

125. The following section occurs among the Paisley beds, and is the more interesting, as it exhibits both shell deposits; the order is descending:—

1. Alluvial soil, 4 feet.

2. Sand and gravel, with *Cardium edule*, other littoral shells in great abundance, and wood bored by *teredo*, 9 inches.

3. Clay without shells, 6 feet.

4. Fine clay, without stones, with *Tellina proxima, Leda oblonga*, and other Arctic species.

5. Laminated clay, without shells.

The boulder-clay does not appear in the section, but is seen not far off, and no doubt underlies No. 5. This is, therefore, a very typical section, although the measurements vary even in different parts of the same pit.

The following is a list of the shells from the clay-bed No. 4, supplied by the Rev. H. W. Crosskey, but not all collected in one pit:—

Cylichna cylindracea.
Mangelia pyramidalis.
Trophon scalariformis.
——— clathratus.
Fusus antiquus.
Buccinum undatum.
Natica clausa.
——— Grœnlandica.
Lacuna vincta.
Littorina littorea.
——— littoralis.
——— rudis.
——— neritoides.
Puncturella Noachina.
Patella vulgata.
Anomia ephippium.
——— aculeata.
Pecten Islandicus.
Leda oblonga.
—— pygmæa.
Nucula nitida.
——— nucleus.
Modiola modiolus.
Mytilus edulis.
Lucina flexuosa.
Cardium exiguum.
Astarte elliptica.
——— compressa.
Cyprina Islandica.
Tellina proxima.
Panopæa Norvegica.
Mya truncata.
Saxicava Arctica.
Balanus crenatus (?).
——— balanoides.
——— porcatus.
Serpula vermicularis.
Echinus—species extinct in the Clyde.
Several species of Crustacea, Entomostraca, and Foramenifera, the species not yet determined.

The shell clay contains stones, rounded and smoothed, and some, though rarely, striated. The beds above the shell clay also contain stones; of the entire number found, the great majority are local; a few are such as could only have come from the north-west.

126. A good section occurs at the tile works in Kilchattan bay, Bute. It is given by Mr Geikie as follows, in descending order:—

1. Vegetable soil.
2. Sand and gravel, stratified and passing downwards into a sandy clay, with gravel, 10 or 12 feet.
3. Red clay (olive green below, without stones or shells), 1 to 2 feet.
4. Fine dark clay, with Arctic shells, 2 feet.
5. Fine laminated clay, red or brown, without shells or stones, 15 to 18 feet.
6. Boulder-clay, with striated stones; the upper surface hummocky and irregular.

Tellina proxima abounds in the shell bed. Mr Crosskey's collection from this bed contains also *Mya truncata, Natica clausa, Leda oblonga, L. pygmæa, L. candata, Syndosmya alba,* with fragments of two species of *Echinus.* There are also indications of an upper shell bed.

Another bed worthy of notice occurs at Lochgilphead. The locality is the bank of a stream traversing the low ground between Lochgilphead and the Crinan canal, and is little raised above the sea level. The section is as follows, order descending:—

1. Vegetable soil.
2. Gravel, coloured by iron, 1 to 2 feet.
3. Bluish-grey sand, 2 to 3 feet.
4. The Arctic shell bed, bluish-grey clay.
5. Boulder-clay, with striated stones, resting on highly-polished rock surfaces.

The laminated clay, so persistent in the Clyde sections, is here absent, and the shell bed, as in Arran, rests on the boulder-clay, filling up hollows in its undulating upper surface. The bed is also remarkable as shewing an extremely quiet deposit. Large and numerous specimens of *Mya truncata* and *M. ude-*

vallensis occur in vertical rows in the clay, in the position in which they bored; the clay which fills the shell still contains the syphon in a perfect state. The deposit is characteristically Arctic. It was fully described by the Rev. H. W. Crosskey in a paper laid before the Philosophical Society of Glasgow, and the following species were enumerated:—

Acmæa virginea.
Buccinum undatum.
Cylichna cylindracea.
Fusus antiquus.
Littorina littorea.
Nassa incrassata.
Natica clausa.
Puncturella Noachina.
Rissoa ulvæ.
——— rubra.
Trochus cinerarius.
———- tumidus.
Trophon clathratus.
———- scalariformis.
Purpura lapillus.
Lacuna vincta.
Anomia ephippium.
Astarte borealis (double valves).
——— compressa.
——— elliptica.
Cardium exignum.
——— edule.
——— echinatum.
Cyprina Islandica.
Leda pernula.
Lucina flexuosa.
Modiola modiolus.
Mya truncata.
—— udevallensis.
Pecten Islandicus.
Saxicava Arctica (in natural boring position).
Syndosmya alba.
Tellina proxima.
Thracia phaseolina.
Balanus udevallensis.
Serpula vermicularis.
Fragments of Crustacea, with numerous Foramenifera and Entomostraca.

Fossils in Boulder-Clay.

127. It has been already stated that no case of fossils in or under the boulder-clay has come under our notice, either in Arran or elsewhere, and the causes likely to mislead observers have been pointed out (Art. 85); still we have the authority of several competent observers for the occurrence of fossils in this deposit. In the year 1820 Mr Bald, the eminent engineer, found an elephant's tusk in the excavation for the Union Canal, 28 miles from Glasgow.* A bank of clay was undermined, and fell into the cutting, when a tusk 39 inches long,

* Wer. Mem., iv., 58.

13 inches in circumference, and $25\frac{3}{4}$ lbs. in weight, was found among the earth which fell. Here it is necessary to observe that Mr Bald divided the superficial deposits into two groups—the recent alluvial cover, corresponding to our upper drifts, and the old alluvial cover, or till, with which he classed sand, gravel, and plastic clays, over what we have called in this account the true till. Now, Mr Bald says that he did not himself take the tusk out of the earth that fell, but that he satisfied himself it was from the till that it came, and that it could not have been in so perfect a state of preservation had it lain in a sand bed; it must have been a hard clay, impervious to water. It was clearly made out that the depth from the surface could not have been less than 15 feet, and may have been 20 feet. In 1817, two tusks of the elephant, and some small bones, were found at Greenhill, on the Carmel-water, parish of Kilmaurs, $17\frac{1}{2}$ feet from the surface, in a brown clay, which appears to have been the boulder-clay. One tusk fell to pieces; the other measured 3 feet $5\frac{1}{2}$ inches in length, was $13\frac{1}{2}$ inches in circumference, and weighed $20\frac{1}{2}$ lbs.* Mr Bald did not visit this locality, but the account he received from Lord Eglinton's factor led him to believe that the recent alluvial cover, not the till, was the repository of the bones. Dr Scouler and Prof. Wm. Couper visited the place twenty-five years afterwards, and ascertained that nine tusks in all had been found, those of four elephants and a-half; and they both considered that the deposit was the true boulder-clay or till. Dr Scouler found, on this occasion, a fragment of an elephant's molar. Shells were found with the elephant's remains by the Earl's factor; but they were not seen by any one versed in the subject, so that the evidence here fails. Dr Couper brought away some specimens on this visit, and placed them in the Hunterian Museum; and among these Dr Scouler has lately found, as Mr Geikie informs us, portions of the antlers of a reindeer.

* Wer. Mem., iv., 63, and iii., 525; Drift of Scotland, p. 71.

A bone of an elephant was found by Mr John Craig in a stratified bed in the till at Chapelhall, near Airdrie, 350 feet above the sea level.* It is very likely that this was a case of a hollow in a great undulation of the boulder-clay. The height of the bed is remarkable. The stratified beds observed in the boulder-clay by Mr Géikie (Art. 120, sub. fin.) contained only rootlets and fibres of plants, those of marsh plants and an *erica*, and the seeds of a *ulex* or furze.

Mr Smith of Jordanhill states (*Wer. Mem.*, Vol. viii., *Research. in Ter. Geol.*, p. 141) that he once found in the boulder-clay broken and water-worn shelly fragments, among which he was able to make out *Cyprina Islandica* and *Balanus udevallensis.* He does not, however, give a section of the beds.

128. By far the most remarkable case of fossils in the boulder-clay is that recorded by Mr Smith, as occurring between Airdrie and the Monkland Iron-works, about fourteen miles S.E. of Glasgow. Here a bed of stratified clay, said to be in the middle of a mass of till, contained many specimens of the common Arctic species of the Clyde—the *Tellina proxima.* "The shells were discovered by Mr James Russell, an operative miner, in digging a well" upon a surface 524 feet above the sea. "After passing through the till, he came to a bed of brick clay containing the shells, fourteen feet from the surface, and therefore 510 feet above the sea level. I could entertain no doubts as to the nature of the superincumbent matter, as that part of it which had been thrown out was left lying at the mouth of the well; it was unquestionably the true till. Indeed, if I had entertained any doubt on this point, it would have been removed by the discovery of a small granite boulder which was found about two feet above the bottom of the till." . . . .

Looking at this case recorded in 1850 (*Jour. Geol. Soc.*, April 24, 1850), with the experience of fourteen years, we should think it very extraordinary, *a priori*, that the shell *Tellina proxima*—characteristic of the Arctic bed above the boulder-

* Proc. Geol. Soc., iii. 415.

clay, and divided from it in Clydesdale by the bed of laminated clay—should also occur in a bed below it; and should, therefore, be inclined to consider this as a decided case of a stratified bed of Arctic shells occurring in the hollow of an undulation of the boulder-clay. An observer so skilled could hardly be mistaken as to the true nature of this bed, though it was seen only "where thrown out at the mouth of the well." It is quite conceivable, however, that the section made by the well should cut both beds without shewing their true relations.

Fossils of Doubtful Age.

129. It was stated in our two former editions that in the spring of 1855, Dr A. Beveridge, of Glasgow, had found a portion of an elephant's tooth in the boulder-clay near Bishopbriggs. We have now reason to doubt that the deposit was the true boulder-clay; and it is perhaps no longer possible to determine this point, the precise spot being unknown to us, and Dr Beveridge having left the country. The specimen is in the possession of Dr Allen Thomson, Glasgow University, and is in fine preservation.

Antlers of the reindeer were found some years ago in the alluvial clay on the north bank of the Clyde, opposite Jordanhill; but it is not known of what age the clay is, or whether it was the original repository of the bones. A cranium of the extinct ox, *bos primigenius*, was found at the same time. The remains were described by Dr Scouler, *Edin. New Phil. Jour.*, Vol. lii., p. 135.

In 1860, during the progress of the cuttings on the Forth and Clyde Junction Railway, the antler of a reindeer was found associated with several species of Arctic shells, at the height of 103 feet above the sea level. The section consisted of the following beds:—

1. Vegetable soil.
2. Stiff clay, with stones, 12 feet.
3. Blue clay, 7 feet.
4. Sandstone rock.

The antler was dug out of the clay, No. 3, close upon the sandstone rock, 18 feet from the surface; the shells were in a similar position, a few yards' distance. The shells found were:—*Cyprina Islandica*, *Astarte elliptica*, *A. compressa*, *Littorina littorea*, *Fusus antiquus*, and fragments of *Balanus*. The association renders it probable that this blue clay was the Arctic shell bed, and that the boulder-clay was here absent. The locality is in Kilmaronock parish, immediately adjoining Croftamie hamlet, a mile from Endrick water, and four miles from Loch-Lomond. This case and that of Kilmaurs are of very great interest, on account of the altitude and of the association of Arctic shells with the other remains.* It is, however, maintained by some that the reindeer inhabited Scotland down to the twelfth century; yet it must have co-existed with the elephant and the Arctic shells, though surviving into the human period, as proved by the associated remains in the two localities.

We must place here, as of doubtful age, the shelly deposits described by Mr J. Adamson and Capt. Laskey (*Wer. Mem.*, Vol. iv., part 2, pp. 334, 568; 1821–23). Mr Adamson describes three localities on the shores of Loch-Lomond where the shells occur; the first, two miles N.W. of the mouth of the Endrick, 8 or 10 feet above the highest level of the lake; another at the south-east angle of the lake, and between the summer and winter levels; the third on the island of Inch Lonach, opposite Luss, in a similar position. The shells were in brown clay, passing downwards into yellow, and over the clay was a bed of gravel. The following is the list of shells:—

Buccinum reticulatum.	Pecten obsoletus.
Nerita glaucina.	Anomia ephippium.
Tellina tenuis (?).	Balanus communis.
Cardium edule.	——— rugosus.
Venus striatula.	Also,
——— Islandica (Cyprina).	Echinus esculentus.
Nucula rostrata, young.	

* Dr J. A. Smith, *Edin. New Phil. Jour.*, N. S., vii., p. 165, 1857.

The deposit described by Capt. Laskey was found on the line of the Ardrossan canal, four miles from Glasgow, and 40 feet above the sea level, and consisted of a bed of sand and clay, with the following shells:—

Turbo littoreus.	Venus Islandica (Cyprina).
—— rudis.	—— striata.
—— terebra.	—— literata.
Arca minuta.	Balanus communis.
—— nucleus.	Anomia ephippium.
Patella vulgaris.	Tellina plana.
—— pellucida.	Cardium echinatum.
Buccinum undatum.	Nerita littoralis.
—— lapillus.	—— glaucina.
Mytilus edulis.	Mya truncata.
Pecten opercularis.	Trochus crassus.

Features of the time.

130. We are now enabled in some measure to restore the features of the surface at the period in question. The distribution of the shell bed marks the limits of the former sea, and its elevation the amount of the depression, *less* the depth due to the particular species. The greatest recorded elevation is 510 feet, but it is not necessary to suppose that the whole area over which the shell bed prevails was depressed to this amount; local variations might occur, as they are known to do in cases of elevation of the land in historic times. All the facts, however, point to an extensive simultaneous depression, yet only to such an amount that all the great dominant features of the land were nearly the same as now. There was an inner sheltered estuary, and an outer open and stormy margin. This is shewn by the absence of the bed of laminated clay in Arran and at Lochgilphead, and the broken state of the Arran shells. Clydesdale, in fact, formed a gulf sheltered by the Lennox hills and coast range of Renfrew, and was connected to the outer waters by a strait below Erskine, and a long narrow channel in the direction of the Dalry valley.

Bute and the connected isles would form, for a long period, a low archipelago, sheltering the inner frith; Bute itself consisting of four islets. Cantire was crossed by several channels, admitting the ocean—one of them debouching upon Lochgilphead, in the line of the Crinan canal. But then, as now, the mountains of Arran and the Cowal district formed a great outer barrier against the access of the ocean. Thus, throughout all the inner waters of the frith of that time, conditions existed favourable to the development of beds of fine sediment which the stormy waters and tidal currents, sweeping the south coast of Arran, would not allow to settle down over the surface of the boulder-clay when the land was sinking. The origin of this stoneless, unfossiliferous clay, is very difficult to understand. We have already traced, in speaking of the Arran drifts, the probable origin of the other beds; the origin of the laminated clay is now the most difficult part of this great problem. If deposited in the sea bottom, by currents passing in from the outer ocean under the frozen surface of the inner frith, the total absence of fossils is difficult to understand, though the absence of stones would be explained; whereas, if formed from mud drifted off from the ends of glaciers by currents passing along shore, or dropped from ice floating about in the unfrozen frith with a load of earth and stones, the absence of stones from the clay is equally difficult to explain. The fineness and perfect lamination of the clay, and the total absence of sand and gravel, indicate a slow and tranquil deposit, far out from the shore, in a deep sea with access to the ocean, but either frozen over or free from floating ice. Are we to suppose it so deep that no shells could exist, and that an enormous elevation of the bottom took place before the deposit of the Arctic shell bed began over it? But this supposition involves great difficulties. Difficulties, indeed, beset every supposition; and we can only wait for more extended observation of the conditions under which it is associated with the other beds, and a strict definition of its geographical limits. We shall then have grounds for a safe generalisation. Meanwhile we may note the great significancy of the contrasts we have stated, in regard to any theory of its origin. It is persistent in Clydesdale and the sheltered shores of the frith; and it exists at Errol, in Perthshire, an inland sheltered situation; but it is absent on the shores of Aberdeen; at Lochgilphead, a situation once exposed, on an open frith; and is not found on the exposed shores of Arran, where also the state of the shells in the bed that takes its place is very different from that in which they are found in places where the laminated clay occurs. From this clay to that above it the change is abrupt everywhere. New conditions were suddenly introduced all over the district. Animal life was rapidly and richly developed on the sea bottom; on this particular horizon the sediment became less fine, was more rapidly let down, and it was charged in many places, but unequally, with stones, dropped, no doubt, from floating ice. The life period, as indicated by the fauna of the shell bed, is remarkable, and proves clearly the prevalence of a severe climate; but precise data are yet wanting in regard to *successive stages* in this period. Yet are there indications of a greater severity at first, and a gradual amelioration afterwards, during the progress of which those species made their appearance which still remain with us, though rarely found; and the conditions became assimilated to those which now prevail; the place of the Arctic species being gradually occupied by that assemblage which now fills our teeming waters.

Boulders, Striated Rocks, and Roches Moutonnées.

131. The loose surface boulders which are found over Clydesdale consist of rocks which exist only to the north-west and west of Glasgow. These have thus travelled *up the river basin* in a direction contrary to that of the present drainage, or action of that force which might in some cases be conceived adequate to the transport of at least the smaller ones. They cover all rock formations alike, being found along the summits of the trap ranges of Lennox and Cathkin, on

both sides of the basin, as well as over its interior and less elevated districts. When we pass westwards of the coal tracts, into the district where Devonian and primary rocks prevail, we find that the transport has not been mutual; that region has received no boulders from the basin of the Clyde, though the limestones, sandstones, and other "measures," with the trap rocks overlying, are sufficiently hard and coherent to have withstood the attrition of a lengthened journey.

132. Grooved and scratched rocks were observed near Glasgow by Colonel Imrie in 1812 (*Wer. Mem.*, ii., 36), and ascribed by him, and by subsequent observers, to the action of the rocks on one another during transport by currents of water sweeping the surface; and were classed under the head of Diluvial phenomena, being supposed to have been produced by the Deluge, or several successive floods in different periods. About 1840, however, they began to be referred to the action of glaciers, or the agency of floating masses of ice, which carried the boulders, grinding them down and marking the surfaces across which they passed. Mr Smith's remarkable discovery of Arctic species in the Clyde shelly deposits is strongly corroborative of this latter view, and indicates a prevalent low temperature immediately preceding that condition of things during which the British seas were furnished with their existing testacea. To this view the opinions of geologists now chiefly lean; but the subject has yet many difficulties. It is impossible for us as yet to pronounce with certainty on the origin of these superficial beds. We may suppose that great undulations in the waters under which the coal strata were deposited produced by earthquakes or other elevating forces, which gradually converted the area into dry land, were of such energy and long continuance as to tear up and re-arrange the rocky materials on the bottom—to sweep off great bodies of strata, forming such valleys of denudation as that of Campsie and Lennoxtown—on opposite sides of which the strata exactly correspond—and that the deposits so formed were subsequently modified by the action of floating

ice, or the surface moulded by glaciers after the whole area was converted into dry land. Icebergs, we know, are only formed along the sea borders of land covered with glaciers. Such ice-covered land must, on this supposition, have existed along the north-west of Scotland, presenting a bold coast towards the south-east, whence the bergs, floating off through a deep sea, would deposit their rocky load as they melted below and were capsized, or as they stranded on the tops of sunken ridges. Thus huge masses of mica slate might be left on the summits of the Pentlands and Lennox hills, and the whole basin of the Clyde be strewed with the spoils of the Grampians, to whose eastern flanks, as an old sea border, we may thus track back the course of the floating bergs. The singular grouping of the blocks favours the view of their being left down on the stranding of successive icebergs. We have seen, in many of the glens descending upon Loch-Long from the west and north-west, successive assemblages of granite blocks, as many as 200 within a circular area of 30 yards diameter, with long intervening spaces almost destitute of them. Other facts favour the theory of former glaciers and enveloping ice-sheets. The masses increase in magnitude as the granite nucleus whence they emanated is approached; and as the glens become narrower, and the hill sides steeper, the crowd of blocks is so great as almost to fill the valley, while the rocky sides exhibit grooved and polished surfaces and *roches moutonnées* in great perfection. In the same district are many striking examples of perched blocks. At heights varying from 1,200 to 1,800, and perhaps 2,000 feet, and on the slopes of mica schist mountains so steep that one cannot descend but by a zig-zag course, granite blocks are lodged but slightly in the thin soil, on narrow ledges or terraces, in considerable numbers and of all sizes, many as much as from one to two tons in weight. Towards the bottom of the valleys some were estimated at 13 to 15 tons. Across the deep hollow occupied by Loch-Long, up the sides of the hills on the east side, and across the rugged and high ridge dividing it from Loch-

Lomond *trainées* of such blocks may be followed. Near Arrochar, east side of Loch-Long, they are abundant and of great magnitude. A vast amount of submergence during the drift period is thus indicated, if the Diluvial theory or that of icebergs is adopted; or, if we adopt the glacial theory, we must suppose a vast thickness of ice filling the valleys, creeping up and over the ridges, and debouching by various openings on the lowland plains.

The shores of Loch-Lomond present many isolated boulders of granite, but few are of such enormous size as above stated. These phenomena, however, come under a large class, belonging to general geology, and need not be further referred to. It may merely be remarked, with reference to the remains of the extinct quadrupeds (Art. 8), that they seem to point to the existence of a warmer climate than the present, and to afford evidence of a kind opposite to that afforded by the Arctic testacea. The state of the remains shews that, though associated with rounded gravel and far transported blocks, they belong to animals which must have lived not far from the places where the bones were found; and that their destruction may have been caused by a sudden submergence, or the advance of icy sheets. But that the climate was warmer does not necessarily follow; since elephantoid remains occur abundantly in North Siberia and Arctic America, in such a perfect state of preservation as to prove that the creatures were fitted for a cold climate, and as everywhere in the British isles these remains occur in association with existing quadrupeds and testaceous species.

133. Examples of scratched and polished rocks may be seen in many places near Glasgow. Perhaps the finest instance within a short distance is the summit level of the pass between the Gareloch and Loch-Long, 600 feet high, and about two miles from the village of Garelochhead, fully described by Mr M'Laren; (*Ed. New Phil. Jour.*, xl. to xlii.) Mica slate rocks are here finely striated in a direction nearly from N.W. to S.E. The rocks on the roadside in several places,

from this point to Garelochhead, exhibit the same markings. They may be seen also at the landing-place at Row, and at several points on both shores of the loch, where they extend under the sea. In these cases, as in those already noticed in Arran, the surfaces directed N.W. are planed off, while those turned towards the S.E. are unaltered. But at none of these points are there any striking examples of polishing, nor of that peculiar "Moutonnée" (rounded and bossy) character which is elsewhere exhibited, as at Parson's Green, near Edinburgh, described by Mr R. Chambers, and Jacob's Wood, near Stavely, Westmoreland, described by the author of these notices; and a case at Ulverston, noticed by Professor Philips; all of which were *under a drift covering*, and as perfect and finely exhibited as any case yet noticed in Switzerland.* A case of striation and rounding, scarcely less remarkable, occurs on the high ground west of Rothesay. The slate and quartz rock on, and in the neighbourhood of, Barone hill (520 feet in height), have been smoothed, polished, and striated in the most remarkable way by a force acting from the north, *up the slope* on that side and down upon the other; the rugged prominences of the strata, which rise towards the north, being worn off, while the surfaces directed the opposite way retain their original rough outlines. The rocks have all acquired under this action the round, bossy, or mamillated character denoted by the term "*roches moutonnées.*" The shores of Loch-Lomond also exhibit tolerable examples. At Rowardennan Inn, between the house and shore, there is a good example of striation and rounded rocks. The surfaces are remarkably worn, rounded, and polished; and the striation about parallel to the axis of the lake. Other cases of striation occur in many places along the shores of the lake. Mr James Thomson, civil engineer, now of Belfast, lately noticed a good case on sandstone rocks from which the till had been removed, about three miles from Glasgow, south-east of the road between Auldhouse and Thornliebank, Pollokshaws, and

* Phil. Mag., Dec., 1850; and Brit. Assoc. Rep., 1850.

due south of the manse, near Auldhouse; the direction was here oblique to the road, or running nearly west. Captain Brickenden has described* striations on the rock of Dunbarton Castle, 150 feet above the river, in the fissure intersecting the rock from north to south. The surface of the sandstones on Craigmaddie moor is striated in many places, the direction being nearly E. and W.; and the trap rocks forming the high ridge between Strathblane and Milngavie are similarly marked, in a direction declining a little S. of E.; while they exhibit also rounded and bossy surfaces in great perfection. Fine examples may be seen in Corrie Glen, about one mile west of Kilsyth. Here a vast accumulation of detrital matter with travelled boulders is exhibited in a natural section on the banks of the West burn, as shown in the annexed cut.

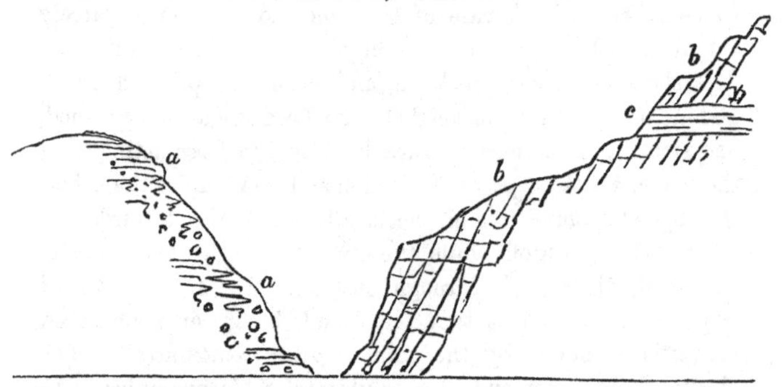

(a a) River bank of detritus, with travelled and striated boulders; (b b) great intruded coulée of basaltic rock, enclosing (c) a shale bed, altered to the state of a coarse opal.

The south bank rises to the height of about 80 feet, and is entirely composed of detritus, partly local and partly derived from rocks existing only *in situ* to the westward. The most remarkable blocks which we noticed were two of large size and rhomboidal form—one of the Campsie "main limestone," and another of greenstone—deeply striated and grooved on one side, that on which they must have rested while borne along upon

* Journ. Geol. Soc., Vol. xi., 1855.

a hard surface. We shall have occasion to refer again to this interesting section.

Moraines.

134. On entering the Argyleshire Highlands from Glasgow, at almost any point, masses of detritus are met with, filling the glens, obstructing their entrance, and backing up against the hill sides in flat terraces, with steep fronts towards the river. The glacialists early fixed on these, and all such accumulations in the Highlands, as moraines of the various orders, terminal, medial, lateral; and not a few have been so described. Mr Charles M'Laren has published an account of a remarkable one in Glen Messan, a glen entering the valley of the Holy Loch from the west, which he considers a true case of an ancient moraine, such being, according to his view, comparatively rare in Scotland.* Mr Robert Chambers, who is somewhat sceptical on the subject of ice as viewed by Mr M'Laren, coincides with him in regarding this glen as a true seat of moraines. Another case given by Mr M'Laren is one mile south of Strachur on Lochfyne. One may be seen at Coruisk, near the mouth of the main glen; another about three miles farther up at Stronlonaig, in a glen entering from the south. The shingle bed at Row Ferry, Roseneath, is fixed upon by Mr M'Laren as the terminal moraine of a glacier which he supposes to have once filled the Gareloch, crossing from the Lochgoil mountains by the col or summit level, where the striated rocks are seen. It may, however, be a true shingle bank formed by the sea, owing to the peculiar movements of the tide at this part. The material is completely sea-worn; and the outlet being narrow, and the extent of water inside great, the tide flows with a powerful current, there being a stream on the west side both at ebb and flood, and an eddy on the eastern. The fact pointed out to me by Mr Smith of Jordanhill, that the shingle bed *rests upon* the boulder-clay, is consistent with

* Edin. New Phil. Jour., Vol. i., New Series, p. 189, 1855; and Vols. xl. and xlii., 1846, 1847.

the idea of its being a moraine, since local glaciers no doubt existed during the period of the upper drifts; but it is difficult to fix upon *the seat* of such a glacier.—It is unnecessary to multiply cases of moraines; they occur in almost every glen as we enter the Alpine country.

III.—CARBONIFEROUS ROCKS.

135. The carboniferous series, on which the beds we have been considering repose, and which attains so vast a development in the basin of the Clyde and adjoining tracts, presents some peculiarities and departures from the normal type, as exhibited in England and Ireland. The entire area occupied by the coal measures is, in the geological sense, but a single basin, with the sole exception of a limited area to the extreme south-east, to be presently noticed. Across the broad zone which reaches from St Andrews and Dunbar on the east, to Ardrossan and the heads of Ayr on the west, the older rocks, on which the coal measures repose, nowhere rise to the surface so as to form "independent basins." Ridges of trap rocks do indeed intersect the area in various directions, but cannot be said to cut it off into distinct basins. The coal tracts of the Forth are continuous with those of Clydesdale, while the latter are confluent with those of Renfrew and Ayr—at either extremity the coal beds rise from beneath the sea. The area is thus a great synclinal trough, filled with an enormous thickness of coal-bearing strata. The permian and triassic rocks are rudimentary, and there is no great body of carboniferous limestone at the base of the series, as in most other countries; and this constitutes a second peculiarity of the Scottish field. The boundaries are formed throughout by Devonian rocks, which attain an enormous development along the north-west border; but on the south-east form a narrow and interrupted zone, so that trap ridges and dikes alone, in many parts, cut off the coal strata from the great southern tract of silurian rocks. Limestone, with all the usual fossils which characterise the carboniferous period, does indeed occur,

but not as a well-developed and continuous base; nor has the millstone grit of England been hitherto established as a member of the Scottish series. It may yet be found possible to prove its existence within the area; and at a still lower level, certain sandstones and shales on the line of junction with the undoubted old red sandstone, may turn out to be on the true horizon of the lower carboniferous limestone. The Ballagan beds seem to us to be such a peculiar group, though by some they may be regarded as the uppermost portion of the old red series. They occur in fine typical development in the bed of the Ballagan burn, at the base of the waterfall called the Spout of Ballagan, about three miles N.W. of Lennoxtown. They here present a vertical section of about 100 feet in height, and consist of numerous alternations of blue and red shales, calcareous marl, white and red thin-bedded sandstones, and thin courses of limestone, there being in all about 230 beds. The lowest bed visible is a coarse, dark-coloured sandstone containing sedge-like plants resembling calamites, but without joints; the highest, a thick-bedded, yellow sandstone without plants, stretching in below the trap series which extends to the hill tops. The plants must, however, exist in the middle portion of the section, for they occur abundantly in the fallen blocks which strew the river bed throughout. A vein of gypsum, 9 inches wide, and numerous contiguous strings of the same substance, stretch far up the cliff, crossing nearly at right angles the various strata, which have a uniform N.W. dip at an angle of 11°. The fossils met with in this locality are these sedge-like plants, which seem all to belong to one species—a species of *lepidodendron*, and a fragment of *stigmaria*. A group of very similar character occurs on the east side of the Leven valley, near Dunbarton, in the Auchinreoch and Dunbuck glens, N. and N.E. of Dunbuck hill; and here a few fish scales have been found, but in so fragmentary a state as to render the species doubtful. The locality is, however, promising, and a careful search for fossils will probably bring species to light which will enable us to fix the exact age of the beds, and determine the true horizon of the carboniferous formations on this side. The strata are almost an exact counterpart of those at Ballagan, but of greater thickness, and seen in more marked superposition to the old red sandstone which here rises northward, forming the outer part of the Kilpatrick hills. This is the western limit of these beds; from Ballagan they sink rapidly eastward, their upper portions being seen in Finglen and Campsie Glen; beyond which they are overlaid by the sandstone forming the floor of the Campsie valley. Beneath the floor of the valley these beds have been met with in borings made by the Campsie Alum Company (*Young, Trans. Glas. Geol. Soc.*, Part I, p. 19). About one mile west of Ballagan they are again seen in the bed of a stream above the village of Strathblane; and still farther north, in Spittal Glen, running up N.W. towards the base of Drumgun hill, they are finely exhibited in several sections, graduating imperceptibly into the beds of the true Old Red. Two great faults here traverse both series, and throw down the Ballagan beds quite out of position. At the head of the glen the latter are directly covered by the basaltic rocks forming the hill tops. The old red sandstone, from this point northwards, is continuous with the great band on the Highland frontier. Throughout the entire district of Lennox, from Dunbarton to Stirling, the geological horizon is greatly obscured by the disturbances attendant on a prodigious outburst of igneous rocks, forming all the higher portions of the hills, and descending in some parts in broad streams into the low country at their base. Such a *coulée* of basaltic rock crosses the high ground between Strathblane and Milngavie, cutting right through the coal measures, which are tilted up by it on the west part of Craigmaddie moor, and separating them from the corresponding measures of the Duntocher district, which reappear in the same relative position as in the Campsie valley, in consequence of an immense fold, or dome-shaped arrangement of the strata. This *coulée* is intersected by a tunnel nearly three miles in length on the line of the new Glasgow Water Works. In the neighbourhood of Castlecarry, Kilsyth, and Croy, similar outbursts and streams of igneous matter occur, altering the coal strata, and in some places bearing them up with it, so that the seams are worked on the hill tops, and in anticlinal beds along their flanks. An elevatory movement due to this cause, producing an anticlinal axis in the centre of the valley below the "spout," has given to the Ballagan beds their actual position; at least, it is only by such a supposition that we can explain their situation and their relations to the strata on the south of the valley, or north-west of Craigmaddie moor. A conical hill of prismatic trap, called Dunglass, above 200 feet high, rises from the centre of the valley, immediately below the section which has been described above.

136. The annexed cut (N.W. to S.E), the lower portion of which on either side of the hill (*f*) is partly conjectural, repre-

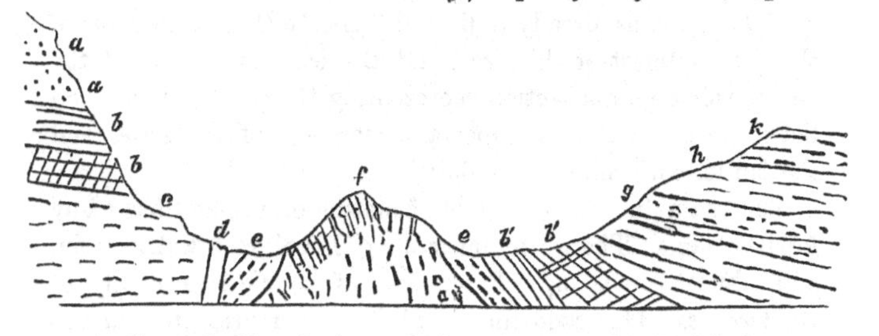

sents the probable relations of the strata at this highly interesting point. Here (*a a*) represents the trap of the hill tops, (*b b*) yellow sandstones, the highest members of the Ballagan series; (*b' b'*) probable position of the same beds south of Dunglass hill; (*c*) the Ballagan beds; (*d*) a trap dike; (*e e*) probable position of the lower portion of the Ballagan series on either side of the axis; (*f*) Dunglass hill, composed of prismatic basalt, and forming the anticlinal axis of the strata and watershed of the valley; (*g h*) coarse gritty sandstones, higher in the series than the yellow sandstones (*b b*); (*k*) thick-bedded, fine-grained sandstones, forming the highest parts of

Craigmaddie moor to the west. The beds of this latter, dipping S.E., form the floor of the Campsie valley, on the level of the Glazert rivulet.

137. The yellow sandstones (*b b*) at the top of the Ballagan section thin out eastward, and disappear in Campsie Glen, so that the trap here rests directly on the beds (*c*), or true Ballagan series, which is depressed in this direction, and covered by newer deposits, whose elevated edges, as they successively crop out N.W., are overlaid along the side of the north hill by tabular masses of trap. The south hill, forming the opposite side of the valley, is a perfect counterpart of the north, save that the overlying trap is absent. The strata correspond exactly; and this rich and beautiful vale, through its extent of five miles in length by half-a-mile to a mile in width, owes its peculiar features to the action of powerful denuding forces, probably coinciding with the elevation of the area, at the close of the glacial period, as already indicated (Art. 132). On account of the great interest of this case, and the economic value of the strata, we subjoin a section representing the principal beds on both sides. The less important members of the series and the faults, with minor irregularities, are omitted.

The main limestone *(e)* is of marine origin, and has many fossil remains. The white limestone *(f)* is of fresh-water origin, and almost made up of remains of *cypridæ*, a family of *crustaceans*. The sandstone *(g)* is that represented in the preceding cut as forming the higher parts of Craigmaddie moor, in consequence of its gradual rise to the west. The alum strata are mentioned farther on.

138. The subdivisions of the coal series, as given in two papers by the late Mr John Craig, mining engineer and geologist, one published in the *Transactions of the Highland Society*, Vol. xii., or Vol. vi., New Series, 1839, the other laid before the British Association in 1840, have not since been disturbed, and are generally received as approximately correct.

The entire series may be divided as follows:—

1. Upper red sandstone series.

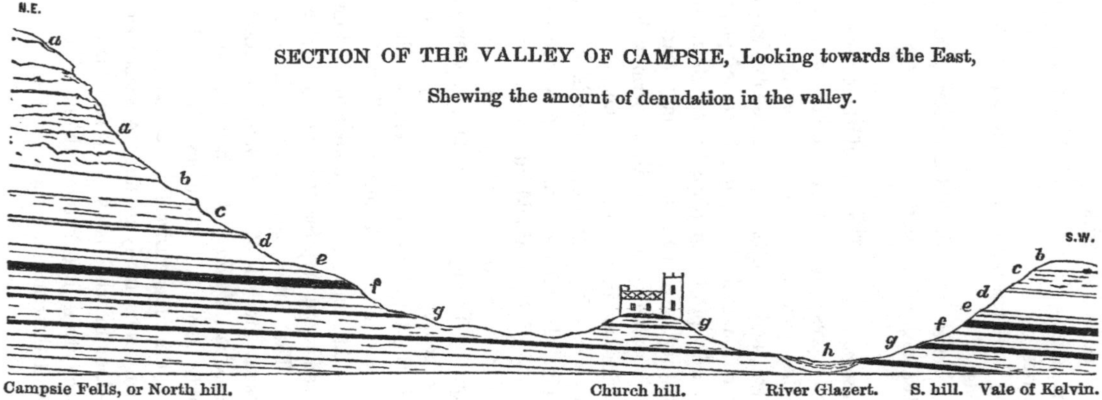

SECTION OF THE VALLEY OF CAMPSIE, Looking towards the East,

Shewing the amount of denudation in the valley.

N.E. S.W.

Campsie Fells, or North hill. Church hill. River Glazert. S. hill. Vale of Kelvin.

(*a a*) *trap of the Campsie hills;* (*b*) *yellow sandstone, underlying the trap, and forming the highest part of the South hill;* (*c*) *shale, with clay ironstone, and seams of limestone* (16½ *fathoms thick*); (*d*) *blue shale* (22 *fathoms*); (*e*) *main limestone* (4½ *feet, marine*), *alum schist* (2 *feet*), *and coal* (3 *feet* 8 *inches*); (*f*) *white limestone, estuary, Kingle, shale, blue limestone, thin coal seams, and fire clay;* (*g*) *thick-bedded white sandstone, forming the floor of the valley, and rising W. into the western portion of Craigmaddie Moor;* (*h*) *River Glazert.*

2. Upper or fresh-water coal series.
3. Upper marine series.
4. Lower coal series.
5. Lower marine limestone series, with intercalated fresh-water beds.
6. The Ballagan series and old red sandstone, at the base of the entire formation.

No. 1, Upper Sandstone Series.—The first of these divisions has been considered to represent the new red sandstone series. It occurs in the higher part of the central district, about Hamilton, Blantyre, etc., and consists of variegated sandstones and marls, with a few thin coal seams, and traces of plants; and hence it seems hardly proper to place these beds in the new red series; but they have been as yet only partially examined, and it may hereafter appear that the uppermost beds really represent this formation in a rudimentary state.

No. 2, Upper Coal Series.—The upper coal series is of fresh-water or estuary origin. It reaches from Glasgow due east by Garnkirk, north side of New Monkland parish, towards Bathgate. On the south-west, the boundary runs from the trap hills at Dychmont by Drumpeller, Bellshill, Motherwell, and Larkhall to Stonehouse; thence by Carluke, east and north-east towards Bathgate; thus comprehending all the central fields in the parishes of Camnethan, Shotts, Dalserf, Dalzell, with Old and New Monkland. Limestones are absent, and are represented merely by calcareous sandstones. The entire thickness is about 220 fathoms. The testacea are all of fresh-water or estuary genera, as *Unio, Anodon, Mytilus,* etc.; fish remains and the usual coal plants abound. There are nine beds of coal, whose aggregate thickness varies from 24 to 34½ feet, and twenty-five minor seams, rarely passing 14 inches each. The first workable bed, or Ell coal, is 45 to 50 fathoms below the upper red sandstone of the first division; the fifth bed is a splint coal, and below it the remaining seams are less continuous, and of inferior thickness. There are, besides, numerous beds of common clay ironstone and blackband, a variety which contains enough carbonaceous matter to effect its calcination; and this is usually done at the mouth of the pits. The blackbands vary in thickness from 4 to 22 inches. There are three principal bands; the upper, 14 inches thick, is 24 fathoms above the Ell coal, and is worked in Old Monkland parish. Another, the mussel band, the second in descending order, is very remarkable as being almost entirely made up of fresh-water shells, which are seldom found in the clays above and below this ferrugineous band. It is 14 to 22 inches thick, and is worked near Airdrie. This blackband is 16 fathoms below the splint coal. The lowest bed is of the same quality, but lies much lower in the series. With the other beds similar shells are associated; and in Ayrshire the same fresh-water coal series appears, with rich blackband ironstones. The area occupied in Lanarkshire by this upper coal series is about 20 miles long, and from 6 to 15 broad.

139. *No. 3, Upper Marine Series.*—Immediately north of the Necropolis Hill, Glasgow, and along the line of the Monklands Canal, this fresh-water series reposes upon limestones, shales, and sandstones abounding in marine remains, *Orthoceras, Encrinites, Bellerophon, Euomphalus, Nucula, Productus, Spirifier*, etc., and constituting the upper marine group. It contains two seams of coal 12 and 14 inches thick, and irregular beds of ironstone; total thickness, about 600 feet. North of Glasgow, across the high ground between Bishopbriggs and Cumbernauld, the sandstones are of great thickness, and the "measures" are without coal.

No. 4, Lower Coal Series.—Below these beds succeeds the lower coal group, without limestone, but with several blackbands and beds of coal, of which the lowest is a cannel coal, 2 to 3 feet thick. These coals and ironstones are worked north-west of the Kelvin, at North Woodside, Jordanhill, etc. At the bottom of the whole group are nodular clay ironstones, and thin limestones.

No. 5, Lower Marine Limestone Series.—The lowest group yet established reaches from this horizon to the Ballagan beds, and contains marine limestones (main limestone 4½ feet thick), alum shale, and sulphureous coal, 4 to 6 feet thick; below which are other limestones of fresh-water or estuary origin, and coal beds, alternating and intermixed with tufaceous trap, to the bottom of the series. At Hurlet, Duntocher, Campsie, etc., these limestones and shales are well seen; but beyond the Campsie district there is no trace of the fresh-water or estuary limestones; they seem to be a local deposit. The alum works at Campsie, on the south side of the valley of denudation already alluded to, have been long established in connection with the shale bed. These works embrace a greater variety of manufactures than any other in Scotland in this department. The aluminous shale, or "alum till," lies between the sulphureous coal below and main limestone above, and is no more than from 2 to 3 feet thick. There is first (ascending) an aluminous band containing iron pyrites, and passing in some parts of the mines into an imperfect blackband ironstone. Over this is the "gentle slate," or principal aluminous band, 6 or 7 inches thick. The uppermost band, called the "diamond" bed, from being studded with crystals of pyrites, is much poorer in aluminous matter, and is seldom used, except to mix with the first or pyritous band. The main limestone rests on this band in the mines; but westwards, on the south hill, the aluminous strata thin out, and the limestone then reposes upon the sulphureous coal. On the north hill the shale bed is 15 feet thick; but the lower portion only is aluminous, and much poorer in the "ore." The strata, however, are an exact counterpart of those on the south hill. The upper part of the shale on the north hill is charged with multitudes of marine shells.—The chief difference at Hurlet consists in the more equal distribution of the aluminous matter throughout the whole thickness of the shale; but there is over the limestone a thick stratum, called the dough or duff bed, rather poorer in sulphur than the alum schist. The schist here gives, when effloresced, long brittle crystals of sulphate of magnesia, which are rarely met with in the Campsie ore. After the underlying coal is wrought out, the alum schist decomposes under the action of the air, exfoliates and falls down. Its sulphur, by the action of oxygen, combines with the metallic bases, forming sulphates. When decomposition is complete, the mass has the appearance of flock silk. The schist in this state is taken from the coal wastes and lixiviated in stone cisterns. The liquid is then evaporated to the proper density, and receives the portion of muriate or sulphate of potash necessary to its formation into the state of a crystallisable salt. This forms the sulphate of alumina, or alum of commerce. Of late years this salt has been developed by slow combustion in long ridges, coated with the exhausted ore—which is found a quicker process.

140. The extraordinary mass of coal at Quarrelton, near Johnstone, which attains a maximum thickness of 90 feet, but is generally 50 to 60 feet, seems to belong to the lower marine division. It lies in a basin-shaped cavity, less than one mile in diameter, and consists of five distinct seams, separated by thin layers of shale, sandstone, or ironstone. Over the coal is a stratum of sandstone, 24 feet thick, and this is covered by blue basalt, 100 feet thick. This prodigious mass of coal, perhaps the thickest ever found, seems to have originated either in vast quantities of vegetable matter, swept into some sheltered hollow in the original estuary, or in a horizontal displacement, which has thrown several contiguous beds over one another.* The strata are displaced by two faults, one throwing the beds down 30 feet, and the other 50, in a direction nearly vertical.

The inferior marine groups appear to the south, south-east and south-west of the fresh-water series, and descend in the usual alternations to contact with the old red sandstone along the borders in that direction.

141. To the south-east of this border, in Lesmahagow

* See Nicol's Geology of Scotland, p. 99, for an illustrative cut, copied from Williams' Mineral Kingdom, Vol. ii., p. 319.

parish and adjoining tracts, a small coal-field occupies an isolated and uncertain position, cut off on the one hand from the Clyde basin, and on the other from the Ayrshire fields, by ridges of Devonian rocks, amid which igneous products are variously intercalated. The strata of the Glenbuck field rise up over a ridge of 1,000 to 1,200 feet high, pass down into the basin of the Douglas, a tributary of the Clyde, and rest on a narrow band of old red sandstone, which cuts them off from the Lesmahagow field; the latter is separated from the Clyde fields by a similar band to the north-east of the village of Lesmahagow, and stretches out east to near the base of Tinto. The strata are extremely well exhibited in natural sections, and contain a complete suite of fossils of true carboniferous types. "The coal shales, sandstones, and ironstones, afford similar remains in abundance; and there can be therefore no doubt that the coal of this field is of the same age as that of the Clyde basins. It is worthy of remark, that several species of Trilobites occur in the shales and limestones, far up in the series; that a white grit, resting on one of the lower limestones, contains a prodigious quantity of fish remains, and corresponds apparently with the great 'fish bed' of some English fields; and that from one of the middle shales a fossil has been obtained agreeing exactly with the *Serpulites longissimus* of the *Silurian System*, pl. v., fig. 1. The field contains fifteen seams of coal, whose thickness varies from 2 feet to 15 feet, the aggregate amounting to about 65 feet of workable coal. There are black band and clay band ironstones, the principal seam of the former averaging 11 inches; but neither these nor the coal beds have yet been worked to any considerable extent, owing to the greater accessibility of the fields farther down the Clyde." * Here, as in other districts, the alternations of limestone and the other measures already noticed descend to the horizon of the Old Red, on which the whole series reposes, without the intervention of any large body of limestone. The series seems to be the re-

* See a paper on this field by the author in Brit. Assoc. Rep., 1850, p. 77.

presentative of those beds of the Clydesdale system which intervene between the Cowglen limestone and the Hurlet coal.

142. The varied phenomena of these interesting fields may be studied to advantage in many natural sections. The Lesmahagow field, through its whole area, and along its borders, is completely opened up by glens and deep water channels, which exhibit the succession of the strata and many remarkable faults. The banks of the Nethan, Calder, and other streams which pursue rapid courses from the high uplands into the profound synclinal trough traversed by the Clyde downwards from the Falls, and the banks of the Clyde itself, present many instructive sections and most favourable localities for collecting fossils. The fresh-water series, across the central area, is less completely exposed, and is best known from interior workings and boring journals. The marine series is laid more fully open. It reaches out on either side to the base of the hill ranges bounding the area; and these, composed of igneous rocks, erupted through the coal formation, or along its outer margin, bear up the various "measures" on their flanks, and bring them to-day in deep glens and the banks of mountain burns. In the Lennox range we have already mentioned the Auchinreoch and Spittal glens, opening respectively into the Leven and Strathblane valleys; the Ballagan, Finn, and Clachan glens, descending into the vale of Campsie, near Lennoxtown; these expose the subjacent Old Red, the intermediate Ballagan beds, and the lower members of the coal formation. Farther east there is an interesting section of beds somewhat higher in the series—one of the most instructive in the whole Clydesdale basin, and to which the attention of the student cannot be too emphatically directed. We refer to the succession of beds laid open in the middle portion of Corrie glen, one mile west of Kilsyth, as shown in the following cut.

The strata here exposed lie between two streams descending from the high trap ranges to the N.W., and uniting farther down before entering the valley of the Kelvin. Of these the

West burn—the upper portion of which is represented in the sketch of Art. 133—cuts most deeply into the soft shales and sandstones which are finely exhibited in a nearly vertical section. Workings on the economic strata farther favour the researches of the student.

The strata belong to the lower marine series. The yellow sandstone (*n*) at the base is higher than that of the Campsie valley, and the two limestones (*g*) and (*l*), respectively 3 and 5 feet thick, are also superior to the Campsie main limestone. The shale beds (*h*) are from 52 to 60 feet thick. The section is from N.W. to S.E.; and the general dip is S.E. at a small angle. Organic remains abound throughout the whole series of beds, and are readily obtained even from the limestones.

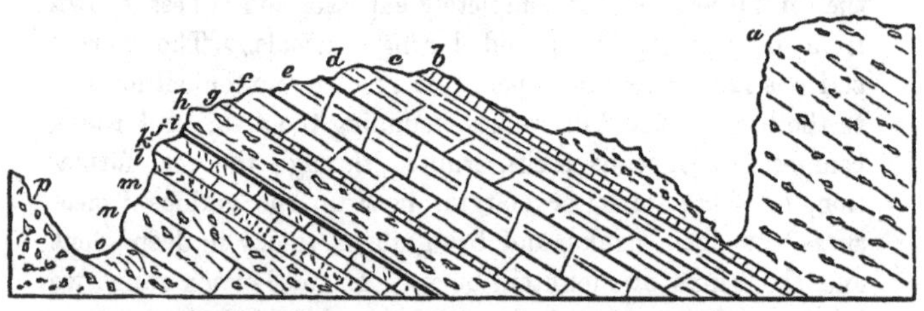

(*a*) *Shale with ironstone layers;* (*b*) *shale with encrinites;* (*c*) *thin-bedded, yellow sandstone;* (*d e f*) *thick beds of calcareous shale with limestone bands;* (*g*) *main limestone;* (*h*) *shale with nodular limestone;* (*i*) *coal bed;* (*j*) *shale;* (*k*) *shale with marine shells and corals;* (*l*) *coralline limestone;* (*m*) *shale with ironstone bands;* (*n*) *thick-bedded, yellow sandstone;* (*o*) *thick shale beds with ironstone bands;* (*p*) *boulder-clay—see p. 294.*

The principal are many beautiful corallines, some very rare; many species of *Productus, Spirifer, Terebratula, Orthoceras, Bellerophon, Goniatites,* etc. The shales contain large *Mityli, Nuculæ,* etc. The remarkable effects of igneous action in this glen will be noticed farther on.

143. We subjoin several journals of borings in these lower marine beds, in order to shew the general correspondence at remote points, as well as the local variations of the deposits. No. 1 gives the strata on the South hill, bounding Campsie valley in that direction; No. 2 the beds lying in the valley, to

be regarded as a continuation of No. 1. They are not presented in any natural section, and seem to lie between the Ballagan beds and the strata of No. 1; the sandstone on the floor of the valley forms the passage from the one series to the other. A portion of the beds in No. 1 is well seen in Craigen glen, opening southwards towards the village of Torrance; and these abound in organic remains, among which, in addition to the genera above mentioned, are *Lingulæ*, and many fish-teeth and scales. The beds here, indeed, are more prolific in organic remains than in any other locality in the district. The boring No. 3 is taken from a pit sunk at the E. end of the South hill, and on the S.E. side of a great greenstone dyke which crosses the district from N.E. to S.W., and throws down the beds on the side of Campsie valley to a considerable depth. They are higher in the series than any in the Campsie district. In the language of the miners, "blaes" denotes all pure shales; "fakes," laminated sandy shales; "kingle," hard siliceous bands; "doggar" beds are bands of shale with nodular clay ironstone.

No. 1.—SECTION OF STRATA PASSED THROUGH IN SINKING A PIT NEAR THE SUMMIT LEVEL OF THE SOUTH HILL, 450 FEET ABOVE THE LEVEL OF THE RIVER GLAZERT.

	Fath.	Ft.	In.
Sandstone, out of section, thickness not known.			
Blue shale,	11	1	0
Ironstone band,	0	0	7
Blue shale,	3	2	2
Black limestone,	1	0	0
Shale, with 4 bands ironstone,	1	0	0
Blue shale,	22	2	0
Limestone (marine),	0	4	0
Alum schist,	0	2	0
Coal,	0	3	8
Under clay,	0	0	7
White limestone (freshwater),	0	4	6
Kingle,	0	4	6
Shale,	1	1	4
Limestone (marine),	0	0	10
Fakes,	0	0	10
Alum schist,	0	0	9
Parrot Coal,	0	0	4
Common Coal,	0	0	10
Fire-clay, in 2 beds, parted by kingle,	1	2	6
	46	2	5

No. 2.—Journal of Bore put down at Burnhouse near the level of the Glazert.

	Fath.	Ft.	In.
Surface earth, ...	1	0	7
Sandstone, ...	1	0	9
Very hard band, . .	0	1	0
Fire-clay, ...	0	1	8½
Fakes, ...	0	1	8
White kingle, ...	0	4	6
Hard shale, ...	1	5	8
Fakes, ...	0	1	6
Fire-clay, ...	0	0	11
Impure limestone, . .	0	3	9
Shale, with stripes of fakes,	1	4	6
Blue shale, ...	0	1	7½
Ironstone, ...	0	0	3
Shale, ...	0	0	7
Coal, ...	0	0	11
Fire-clay, ...	0	1	2½
Fakes, ...	1	0	6½
Kingle, ...	0	0	7
Limestone, ...	0	0	8½
Shale, with stripes of fakes,	1	4	10½
Shale, ...	0	1	1
Blue fakes, ...	0	0	7½
Shale, ...	0	1	1½
Ironstone, ...	0	0	3
Shale, ...	0	0	4
Coal, ...	0	1	2
Fire-clay, ...	0	2	1½
Fakes, ...	1	0	10½
Impure limestone, . .	0	1	10½
Shale, ...	1	5	5
White kingle, ...	0	2	10½
Hard shale, ...	0	3	6½
White freestone, . . .	0	2	5½
Kingle, ...	1	0	7
Gray fakes, ...	0	1	7
White freestone, . . .	2	1	9½
Fire-clay, ...	0	2	9
Light fakes, ...	0	2	0½
Kingle, ...	0	1	5
Fakes, ...	0	1	0
White kingle, ...	2	4	6
White marble band, . .	1	1	11½
Fire-clay, ...	0	1	5½
Blue shale, ...	0	0	2½
Light shale, ...	0	1	3½
Shale, with stripes of fakes,	0	1	6
Kingle, ...	0	1	7½
Fakes, ...	0	0	6
Shale, ...	1	0	0
Shale, with stripes of fakes,	0	0	9
Rock not known, . . .	0	0	9
Limestone, ...	0	1	8
White freestone, . . .	0	1	5
Light fakes, ...	0	1	5
Kingle, ...	0	0	4
Light fakes, ...	0	0	9½
Kingle, ...	1	4	9½
Light fakes, ...	0	1	6
Light kingle, ...	0	3	9½
Depth of this bore below the level of the valley, .	33	0	0

No. 3.—Journal of Strata passed through in Sinking a Pit for Coal on the Grounds of Balquarhage, near Campsie.

	Feet.
Surface earth, with clay and boulders,	7
Blue shale, with marine shells, .	42
Shale, with encrinites and trilobites,	3
Shale passing into limestone, .	1
Culmy limestone, in two beds, .	8
Calcareous shale, ...	1
Bituminous shale, ...	2
Dark blue shale, ...	2
Shale, with ironstone nodules, .	2
Dark soft shale, ...	1
Black shale, with doggars, . .	1½
Bituminous shale, ...	0½
Coal, not very good, ...	5
Dark sandy shale, full of coal-measure plants, ...	3
	79

No. 4.—Economic Strata of Lanarkshire, copied from Section made out by Dr Rankine, and Mr John Kerr of Castlehill.

	Fathoms under New Red Sandstone.	
Upper or fresh-water series.	32	Upper coal.
	41	Cone in cone limestone.
	52	Ell coal.
	56½	Fresh-water limestone.
	63	Pyotshaw coal.

	Fathoms under New Red Sandstone.	
Upper or fresh-water series.	64½	Main coal.
	78	Splint coal, with band of clay ironstone above it.
	86	Mussel band ironstone.
	88	Sour milk coal.
	93½	1st Black band ironstone.
	108	Virtue-Well coal.
	114	2d Black band ironstone.
	118	Mussel band ironstone.
	124	Castle hill 1st coal.
	133	Castle hill 2d coal and ironstone.
	144	Kiltongue coal.
	164½	Drumgray coal.
	177	Coal (*no name given in section*).
	190	Crofthead coal and ironstone.
	200	Slaty ironstone.
	220	Thomson's ball and gin ironstone.
	235	Curdley ironstone.
	254	Gare limestone.
	282	Belstone limestone.
	308	Climpey limestone.
	325	Carluke 1st coal.
	326¾	Carluke 2d coal.
	335½	Tower coal.
	342	Four bands clay ironstone.
	347	Gas coal and black band ironstone.
	370	Lingula ironstone.
	380	Lingula limestone.
	381½	1st Kinshaw limestone.
	384½	2d Kinshaw limestone.
	390	1st Culmy limestone.
	393 to 397	Rae's gill ironstone, 11 bands.
	399	Hosie's limestone.
	415	2d Culmy limestone.
	418	Main limestone, under a band of clay ironstone.
	419	Coal under main limestone.
	424	Limestone.
	434	Shelly limestone.
	440	Oyster limestone.
	450	Scabricula ironstone.
	465	Old red sandstone.

A Table drawn up by Mr William Moore, civil and mining engineer, will be given farther on, containing all the economic strata of the several fields, as they exist at the present time, in their true relative positions.

The preceding section, No. 4, contains the economic strata only of the upper portion of Lanarkshire. It was prepared for the meeting of the British Association in 1855 by Dr Rankine of Carluke, long well known for his profound and accurate acquaintance with these strata and their fossils. The first 235 fathoms give the fresh-water series; below this, from the Gare limestone down to the oyster limestone, inclusive, there are thirteen beds of marine limestone: and among these, the beds 418 and 419 are the equivalents of the Campsie main limestone and coal; and the Campsie strata generally are represented by the beds from the Kinshaw limestone (381½), inclusive, downwards to the old red sandstone. In the Campsie district, all the beds above the Kinshaw limestone are wanting; and the Ballagan beds have no representatives in the Carluke district. The Ballagan series does, however, occur in the S. E. coal field on the Merse of Berwick, as we are informed by Mr Stevenson, and indications of them have also been noticed on the borders of the East-Lothian field.

144. Sections very similar to the preceding are obtained around the borders of the district, amid the series of lower strata; but it is impossible to identify the beds except in a very general way, as great analogous groups of limestones, shales, ironstones, and coals; and viewed in this way the Carluke series, so carefully made out by Dr Rankine, is typical of the entire lower deposits. The lower coals are generally of inferior quality, but good ironstones descend to the very base; while limestones, usually marine, occur in oft-repeated alternations with the other measures. A general resemblance exists, but the local exceptions are remarkable. The Campsie, Duntocher, Carluke, and Hurlet series have been referred to already as illustrating this similarity. At Whiteinch, near Partick, a little above the level of the Clyde, a bore of nearly 300 feet gave eleven beds of limestone, varying in thickness from 8 inches to 8 feet, aggregate about 30 feet; eight bands of ironstone, in all about 5 feet, but no coal seams. At Maxwelton, near Paisley, a bore of 200 feet gave

neither limestone nor ironstone, but several good coal seams, aggregate thickness about 14 feet. At Goldilee, near Houston, on the outer margin of the Renfrewshire fields, where they come against the trap associated with the old red sandstone of the Coast Range, among the strata passed through in a shaft of 110 feet deep were two limestones 8 feet apart, and respectively 7 and 4 feet thick, and four coal seams about 6 feet in all, but no ironstone. At Kaimshill, on the same border, a little south of Lochwinnoch, the limestones and coals are similar, but good ironstone bands are associated. At Howwood, near Castle Semple, 125 feet of shales and rich ironstones overlie a bed of limestone 10 feet thick; and three miles S.E., at Braidstane, near Beith, there occurs the largest body of limestone in the west of Scotland. The following section shows the order and thickness of the strata:—

	Feet.		Feet.
1. Surface soil, ...	1	6. Coal, ...	1
2. Limestone, full of marine fossils,	14	7. Slate clay, ...	6
3. Fire-clay, ...	4	8. Limestone, with few fossils, .	18
4. Limestone, ...	1	9. Coal, ...	1¼
5. Clay indurated, ...	1	10. Slate clay, not passed through,	?

The strata are traceable for a mile in the line of bearing, but the breadth is limited in the direction of the dip. They crop out towards the north in the direction of the great outburst of trap, already noticed as forming a mineral axis in this region, in connection with the old red sandstone, through which it has been erupted.

145. On the subject of these local variations, and the prevailing similarity among the local groups, Mr Montgomery of Cloak has the following remarks respecting the fields of Renfrew and North Ayr, in his clear and accurate outline of the geology of these tracts:*—"The different strata alternate in no definite order, while no one of them seems essentially necessary to the formation of a coal-field. Even coal itself may be awanting, whilst all the other strata usually found in a coal-basin are there. The place usually occupied by freestone or slate clay is occasionally filled by ironstone. Limestone is as

* Jour. of Agricul. and Trans. of Highland Soc., Vol. xliii., p. 441, 1839.

capricious in its presence or absence as any other of the measures; nor is it possible even to define its limits in the district under consideration. It makes its appearance here and there, more particularly upon the edges of the coal-fields; but its continuity from one place to another where it is seen is by no means certain. In various places it has been wrought out, perhaps indicating that the quarries had been opened in masses detached by some convulsion from greater strata. But whatever hypothesis may be adopted regarding this matter, none favourable to the opinion of the limestone strata holding a necessary place in the coal measures of Renfrew and Ayr can be entertained. If the strata of limestone were ever continuous, they certainly are not so now; and to support the opinion of their former continuity, a much more intimate acquaintance with their numerous organic remains would be necessary, than it is supposed that any one has yet acquired."

146. We do not know that these remarks, penned twenty-five years ago, need any modification at the present time, with all the knowledge we have since acquired respecting the organic remains of these beds. They support the opinion already advanced, upon evidence furnished by other districts of our great coal tract, that there is no great persistent body of "carboniferous limestone" separating by a well-marked horizon the "upper" from the "lower" coal measures, and that the millstone grit also finds no place in our series. In England and Ireland the development is more complete, and the lines of division more strongly marked. The mountain limestone is an important formation, and widely persistent in both countries; and in England the millstone grit, overlying the limestone, in most cases fully developed. In Ireland it is less universal, and its thickness is much less. In Scotland, on the other hand, the millstone grit has not been identified; and as to the mountain limestone, we may almost say that it is, as it were, split up into numerous bands; and strata of shale, coal, ironstone, and sandstone, intercalated among the separated members. The mineral type differs widely; but the organic remains are

identical throughout the two series. They are thus of one age, but formed under different conditions of the terrestrial surface which adjoined. An ancient sea channel filled the space already described (Art. 135) as forming the great synclinal trough in which lie our coal strata. Warm and humid lands stretched far and wide to the north-west and south-east; they nourished a luxuriant vegetation of tropical or sub-tropical species, and were drained by large rivers emptying into this common receptacle. Like the tropical rivers of our own era, these were subject to sudden and violent floods, which swept away the exuberant vegetation, and spread it widely over the bottom of that ancient frith. Sand and mud, mixed with peaty matter produced by decomposition in sheltered situations, were the river sediment, in an incalculable disproportion to the trees and plants carried down. The calcareous matter, suspended in small quantity in these turbid waters, formed, by chemical segregation, subscrystalline beds of limestone of trifling thickness and limited horizontal extent. Under peculiar circumstances existing in particular spots, the purity of the water, and a lengthened period of repose, favoured the growth of a few corals, and the rearing of pigmy coral banks from the bosom of that primæval sea. In that region of the ancient earth where Ireland is now situated, the conditions were widely different: the whole interior, to two-thirds of the present area, was occupied by the waters of an inland sea or gulf, as large as the Aral Lake. Around its shores on all sides rose high ridges of granitic rocks and silurian slates, breached in a few places by narrow straits, uniting it to the ocean outside; and its surface was varied by a few islands of elongated form, less elevated than the bordering land, but formed of rocks of the same old types. The pure and warm waters of this archipelago were tenanted by myriads of rock-building polypi, whose ceaseless toil through long ages reared up from its still depths numerous reefs of coral; while the high temperature of the waters favoured the prolific growth of other coralline species, and the development of chambered cephalopods (*nautilus, orthoceras*) to an enormous size, bearing to the Scottish species somewhat of the same relation as do the Tridacnas of the Eastern seas to the dwarfish bivalves of our northern latitudes. The suspended lime—calcareous sand formed of triturated coral—and the countless myriads of testacea, zoophytes, and other creatures with which the waters teemed, united in forming vast deposits of limestone with a prevailing crystalline structure rarely found in the limestones of this age in Scotland.

Conditions very similar to these prevailed over large tracts in the centre and north of England, where the formation follows the crystalline type, and has less of that mechanical character which marks the Scottish deposits. Yet are the fossils most markedly alike throughout these groups in the three countries, and all are evidently of the same age. We speak, however, rather of the lower groups, the carboniferous limestones of England and Ireland, and the shale and sandstone beds with coal and ironstone seams in Scotland. The upper portion of the Scottish series is, we have seen, of freshwater or estuary origin, and formed under different conditions; while the upper portions of the Irish series perhaps differ in age from both the others, and in amount of development are far less complete.

147. For Fifeshire and the Lothians, Mr Page has recently* given a classification very similar to that set forth by us, but somewhat less minute, and perhaps on that account to be preferred in the present state of our knowledge. Founding his deductions upon his own inquiries and the published reports of Mr C. M'Laren, Sir D. Milne Home, and Mr Cunningham, he arranges the strata in the following order:—

"1. True coal measures—consisting of numerous alternations of coal, shales, sandstones, ironstones, and occasional beds of impure limestone, . . 2,500 ft.

* In his excellent "Advanced Text-Book of Geology," a work remarkable for accurate and clear statement and sober generalisation, and adorned by many eloquent passages.

"2. Several strata of crinoidal and *Productus* limestone, with intervening beds of shale, sandstones, and thin seams of coal, ... 180 ft.

"3. A vast thickness of whitish, fine-grained sandstones, bituminous shales, a few thin seams of coal, mussel-bands or shell-limestone, and fresh-water limestones abounding in cyprides, ... 1,500 ft.

"In this instance there is no development of millstone grit, the whole system resolving itself into upper coal, mountain limestone, and lower coal. . . . . How far these subdivisions may indicate great life-periods, or only portions of one great epoch, has yet to be determined by a more minute and rigorous comparison both of vegetable and animal species,—a task which has hitherto been neglected for the lighter labour of popular description and attractive generalisation."

The second group in this arrangement corresponds to Nos. 3 and 4 in our division (Art. 138); the third corresponds to our fifth, and in it there occur, as with us, intercalated beds of fresh-water origin, consisting chiefly of a mass of cypridæ. Our sixth group, the Ballagan series of beds, finds no representative in the eastern fields, though occurring, as already noticed, in a state of less perfect development in the Merse of Berwick. But to constitute the middle portion into a separate group under the designation of "the mountain limestone," appears to us only calculated to mislead the student, and to give a false impression to geologists regarding the true characters and great distinctive feature of our lower groups. Thick beds of marine limestone occur at the base and in the middle of the series, and contain the same fossils; while at each end of the series, fresh-water strata are intercalated among the prevailing marine groups.* The en-

* The marine limestones of the Lennoxtown valley, and white cyprida estuary limestone (*f*), have been already noticed (Art. 137). Mr John Young, of the Hunterian Museum, also finds a fresh-water limestone among the marine beds at Bishopbriggs; and very recently Mr John M'Diarmid of Glasgow found within the

tire series of beds thus forms one great natural group of strata, uniform throughout in the character of its often-repeated alternations of similar bands of rock, and in its organic contents. Its prevailing character is that of a mechanical deposit, the crystalline limestone bearing an incalculably small ratio to the shales, sandstones, and impure earthy limestones. Yet is this entire series the representative of the great crystalline masses at the base of the coal formation in England and Ireland. The whole group seems to us to be in the place of the mountain limestone; but this rock nowhere exists as a continuous base or geological horizon. This view was first, we believe, put forward at the Edinburgh meeting of the British Association in 1850, but met with considerable opposition, and has since been often canvassed. Mr Hugh Miller remarked regarding it as follows: *—

"There are few finer sections of the coal deposits anywhere in Britain than those laid open along the shores of Granton, Musselburgh, and Prestonpans; and the section of the mountain limestone exposed in the ravine at Dryden, is, as far as I have yet seen, the most extensive in Scotland. By those who hold, as is done by some of the geologists of our Western capital, that this formation is wanting as a base to the Scottish coal field, a visit to this section might be found very instructive. It does not exhibit that great thickness of limestone for which the corresponding formation in England is so remarkable, but presents for several hundred feet together, in its encrinal bands, intercalated amid shales and sandstone, evidence of a marine origin; and its upper calcareous beds, laden with spirifers and producta, and of very considerable thickness, show that a tolerably profound sea must have

city, south side of Parliamentary Road, opposite the Town's Hospital, beds of limestone and calcareous sandstone, with numerous marine fossils of the ordinary genera, and fossil plants, at the very top of the marine series, and closely adjoining the fresh-water series of our second division (Art. 188).

† "The Fossiliferous Deposits of Scotland," being an Address to the Royal Physical Society of Edinburgh, delivered November 22d, 1854. Pp. 16, 17.

covered the field shortly ere the formation of our older beds of workable coal."

These remarks, however, hardly impugn the statements already put forward. This limestone is not at the base; its thickness is not great; and its encrinal bands are intercalated among shales and sandstones, which contain the same fossils. The sea was indeed deep; but it was not a sea of coral reefs, nor of pure chemical deposits; its waters, by means of frequent floods, were widely discoloured by vast irruptions of black mud, and red or white sand, with which calcareous matter was sparingly mixed; and thus the whole deposit is of one age, while the limestone is truly, from its fossils and mineral character, the same as the mountain limestone, though not *en masse* in the same position. *The shales and sandstones, with its separated portions intercalated among them, represent in all particulars the great English formation, underlying the coal.* —What a long series of changes and vast lapse of time are implied in the elaboration of these strata in the bosom of the primæval ocean! Its deep bed was often laid dry, and on the desiccated surface lakes were formed, which became the abode of cyprides and other fresh-water genera. These conditions so long prevailed, and the individuals continued to exist in such countless myriads, that thick and widely continuous beds of limestone were accumulated from their exuviæ. Again, the area subsided to a small depth, the sea found an entrance, and a thick coal seam and shale beds with marine remains were deposited. A continued subsidence to an enormous amount succeeded, and a vast series of shales and sandstones, and a few seams of coal and beds of limestone, charged throughout with a prodigious variety of marine fossils, were slowly accumulated. Again, by some mighty internal convulsion, a new aspect was given to the region, most probably that of a wide archipelago, in whose winding channels and on its little elevated lands, was formed a new series of alternating marine and fresh-water strata, almost a repetition of those far down in the series. To explain the origin of the upper portion of the marine series, the thick body of strata ranging north of Glasgow and across the higher portions of the city, we must again suppose the existence of a sea of considerable depth, receiving an abundant and very mixed sediment, and the waste of a prolific land vegetation of a tropical or sub-tropical character. The close of this period was marked by the elevation of large tracts along the margins of the area, and the formation of a lake or estuary filling the central and depressed portion, now occupied by the upper or fresh-water coal series already described in Art. 138. The climate remained unchanged, and the vegetation then attained its maximum of development, resembling more that of a tropical jungle or river-delta, than anything now seen in our latitudes. Then began that remarkable series of changes in repeated depressions and elevations of the bottom and margins of this lake or estuary, which geologists have often attempted to chronicle, in describing the physical geography of the coal period. These are much the same in Clydesdale as in other coal tracts, and we need not here attempt the description. We may observe, however, that the under clay, or floor of the coal, *generally* exists in our fields, roots being often seen passing down through it; and that ranges of upright trees have been met with at several levels amid our coal measures. These under clays and seats of the trees point out the successive surfaces on which the vegetation flourished, and which, though perhaps slightly depressed, cannot have been far below water; the coal seams are a sort of rude chronometer indicating the time occupied in the deposit, while the interposed sandstones and shales show the amount of depression which took place before the surface was again fitted for the growth of plants. This series of changes was continued through vast periods of time, till at length the rich storehouse was furnished with those materials which were destined, in the providence of the Creator's goodness, to contribute to the wants and happiness of man, when he should be called into being upon that theatre, already the scene of the creation and extinction of many races which had served their purpose and disappeared. But the series of revolutions was not yet completed. The entire area, with portions of its borders, was again submerged to a depth probably exceeding 300 fathoms; the strata were fractured and dislocated in all directions, and successive streams of volcanic matter were poured out from the heated interior. These spread out among the beds in sheets, pierced them through in dikes, and were accumulated over them in great mountain masses. The succeeding revolutions by which the existing aspects were given to the surface have been briefly sketched in the preceding part of these notices.

148. With regard to the distribution of these igneous products, it is remarkable that they have been chiefly erupted along the boundary of the Devonian and carboniferous rocks, from Ardrossan to the mouth of the Eden, in the Coast Range, the Lennox, Ochill, and Sidlaw hills, so nearly continuous with one another, as to be separated only by the channels of the Clyde, Forth, and Tay. This line is parallel to the principal axis of the Grampians, which were elevated at an earlier period; and is also in the direction followed by the many earthquake movements recently experienced in Scotland. There is also evidence of elevatory movements in connection with another band of these rocks near the border of the primary Highland tracts.* On the south border of the coal districts there is also an extensive development of these rocks; and they constitute several considerable ranges within the area, as the Fereneze and Cathkin hills separating the Ayrshire fields from those of Lanark and Renfrew, the Bathgate hills, and other lesser ridges. The huge trappean mass of Tinto, and the varied igneous products of the Pentlands, seem to belong to an older era. Those of the Campsie hills are referred by some geologists to the age of the old red sandstone. We contend, on the other hand, that the trap ranges on the borders, as well as those within the area, were erupted since

* Jour. Geol. Soc., 1852.

the deposition of the coal strata. Along the base of the Cathkin hills, in the glens near Barrhead and Neilston, and in natural sections among the Gleniffer hills, the coal strata may be seen gradually raised to a high dip, set on end, and finally reversed when in contact with the trap. The Coast Range of old red sandstone from Ardrossan to Port-Glasgow, through which the Clyde has forced a passage between Bishopton and Dunbuck is overlaid throughout by trap rocks. These cover also the eastern slopes, and come in contact with the coal strata, which are overtopped by and disappear beneath them. Similar phenomena have been already noticed in describing the Campsie district. And, besides, had the trap rocks there been erupted and widely spread over the surface of the old red beds before the coal period, trap fragments might be expected to occur in the Campsie conglomerates; but no such imbedded fragment has ever been found. The case figured in the cut of Art. 135 is very interesting. The great stream of basaltic matter intruded into the coal strata, encloses a shale bed *(c)*, which has been altered through a considerable distance to the state of a coarse opal. An impure limestone also in the river bed adjoining contains crystals of galena in close proximity to the trap. In the overlying trap there is a vein of heavy spar with traces of copper and silver. The changes generally throughout the district are similar to those often described in connection with the contact of the trap rocks and the sedimentary strata. Some very peculiar effects will be described farther on.

IV.—THE OLD RED SANDSTONE AND ITS LIMESTONES.

149. In addition to what has been already said regarding this rock as a base to the carboniferous formations, we have now only to notice those portions of it which appear upon the shores of the frith. Those on the north shore are prolongations of the Kilpatrick hills, which, subsiding at the valley of the

Leven, rise again to much lower altitudes in a ridge bounding the Clyde on the north from Dunbarton to the Gareloch. Thence the old red sandstone passes into the peninsula of Roseneath, in which its junction with the underlying old slates is strikingly marked on the features of the landscape. * It crosses nearly through the middle of the remarkable and very picturesque dell, which intersects the peninsula from Campsail bay to Kilcreggan in a direction nearly north-east and south-west. Thus, in external aspect, and in the nature of its rocks, this southern portion is isolated from the rest; it consists of a single hill of a depressed conical form, having a smooth outline, and extending in gentle and fertile slopes to the water's edge on three sides. Here, as in other places, the soil formed by the decomposition of the sandstone contrasts most favourably with that which rests upon the cold retentive clays of the coal formation on the one side, and the old slate rocks on the other. The series exhibits but its lowest members—conglomerates, coarse sandstone, and finely laminated red sand, irregularly disposed. The base of the conglomerate is coarse red sand; and the imbedded fragments are granite, porphyry, quartz, and various kinds of slate; the three former are very much rounded, the latter have lost their angularity, and present elliptic forms. The origin of these is to be looked for in some near district of the Grampians, where such varieties exist, and which we know were elevated and exposed to the action of mechanical forces prior to the epoch of the old red sandstone; the quartz and slate pebbles are from the adjoining strata.

The contact of the sandstone and slate is nowhere seen. In the western part of the cliffs at Portkill bay, the two rocks approach very close, the sandstone dipping at a small angle towards the nearly vertical slate; and in Campsail bay on the east side a considerable space of flat beach intervenes, con-

* It is said that "Roseneath" means, in the Gaelic language, "the little dell, or dingle;" and that the name was given to the whole from this peculiar feature. See *New Statistical Account* under "Roseneath Parish, Dunbartonshire."

cealing the contact. The line of junction passes near the Saw-mill, across the upper part of the fields sloping down from the northern edge of the plantation which crowns the heights on the south side of the great hollow or dingle, and strikes the shore in Portkill bay, near Kilcreggan pier. On the southern coast of Portkill, the sandstone dips for a short distance toward the south; the dip then changes, and continues in other parts to be between west and south-west, at an angle of about 15° to 20°.

About Innellan, and thence to Toward Point, the Old Red with subordinate beds of limestone occupies the coast, occurring in small patches over the slate, extending, however, no farther inland than the well-marked terrace which exists here, whose front shews the slate rock from base to summit. But neither in the sandstone nor limestone have fossils yet been found. Similar limestone bands occur in the Old Red on the opposite coast at Innerkip; and at the south base of Benlomond they descend to the very bottom beds of the formation, not far from the mica slate. In none of these localities are there any fossils.

150. The coast section of Renfrewshire presents only old red sandstone and trap with occasional beds of limestone. The sandstone rises seaward, dipping east and south-east, at a small angle, and everywhere occupies the coast except at Kempoch Point and Cloch Point, where the overlying trap reaches the coast line, and is seen between high and low water, resting upon and altering the sandstone. The great overlying mass of trap sends out innumerable dikes intersecting the sandstone of the coast, and running in very various directions, but with a general tendency to the west and north-west. Into a minute description of these, and the changes produced by them upon the rocks which they intersect, we cannot enter in this place. One instance only will be given, as bearing upon the changes induced on limestone. This rock appears in two places near Innerkip; one bed extends from the bridge on the Greenock road, at the north end of the

village, up into the hill on the south-east of the village, rising with the slope of the sandstone beds, and preserving a thickness throughout of about twelve feet; it has been extensively quarried behind the village, but is now little worked. As it is extremely hard, and contains much chert disseminated in veins and bands, the rock is capable of taking a fine polish, and of being applied to ornamental purposes; the colour is dark gray. Between the limestone and the sandstone above it, there is interposed a bed of loose materials, consisting of red sand, marked with round gray spots, and enclosing pieces of limestone and sandstone. Hence the bed of lime must have been exposed to considerable decomposition before the sandstone was deposited over it. A lengthened and careful search was not rewarded by the discovery of any fossils. It is, however, obviously in the position of the cornstones of England; which have yielded some ichthyolites (*cephalaspis*).

Several beds separated by strata of sandstone occur on the shore at the mouth of the river. The whole series is here traversed by dikes of greenstone, the largest of which is about sixty feet wide, and ranges about N.N.W.; the others are so numerous, and so ramified, as almost to defy description. They pierce through the limestone in every direction, thin veins branching from the greater, and often again uniting, while small portions of the limestone and sandstone are entangled in the trap, and traverse it in disconnected veins. The changes produced upon the limestone are of the most interesting kind; we know of no locality in the west of Scotland where the posterior origin and intrusive character of the trap rocks are so clearly manifested, and would strongly recommend it as a point to be visited by the student of geology. The changes which the limestone has undergone run through every variety of external aspect, from the impure, dark-coloured, perfectly opaque state, to that of a pure white marble, translucent on the edges, homogeneous throughout, and devoid of stratification, or visible lines of cleavage. Intermediate between these extremes there are an indurated semi-crystalline limestone, a granular, saccharine marble, crumbling into fine powder under slight pressure, and phosphorescing when thrown upon a heated surface; a very hard white or blue crystalline marble, having the crystals in distinct plates; the degree of alteration depending on the distance from the sides of the dike. The entangled portions are among those which exhibit the greatest amount of change. The most altered parts of the sandstone resemble quartz rock.

In order to determine whether any and what chemical changes had been induced simultaneously with these alterations of mineral character, and to afford terms of comparison with metamorphic action upon other limestones, I had analyses of several specimens made at the Glasgow College laboratory, under the care of the late Dr Robert Dundas Thomson. These are as follows:—

Specimen No. 1 is the unaltered limestone, as pure a specimen to the eye as could be selected.

No. 2 is a saccharine marble, crumbling readily into powder.

No. 3 is the most altered specimen in this locality, described above as "a pure white marble, translucent on the edges," etc., and which turns out to be table spar.

No. 4 is a carefully-picked specimen of the same, free from carbonate of lime.

No. 5 is a calcareous sandstone, altered by contact, scarcely distinguishable from No. 2.

No. 1.

Carbonic acid,	42·40
Lime,	54·58
Silica,	0·70
Magnesia,	0·40
Water and loss,	1·92
	100·00

No. 2.

Carbonic acid,	42·52
Lime,	54·68
Insoluble siliceous matter,	1·14
Magnesia,	*a trace.*
	98 34

No. 3.—Spec. Grav. 2·88

Silica,	47·50	48·00	46·64	49·38
Lime,	44·90	45·40	46·27	44.59
Protox. of iron and alumina,	5·40	1·61 2·01	1·44	1·49
Water,	0·80	0·80	0·80	1·60
Magnesia,	0·70	0·83	—	1·05
Soda,	—	0·92	—	·94
Carbonic acid and loss,	—	—	4·85	·95
	99·30	99 57	100·00	100·00

No. 4.

Silica,	49·380
Lime,	44·596
Alumina and protox. of iron,	1·490
Soda,	·936
Magnesia,	1·050
Water,	1·600
	99·052

In atoms this is

Silica,	24·89	1·93
Lime,	12·74	1

And the formula is, omitting the impurities—

CaO,2SiO,

or Bisilicate of lime.

No. 5.

Carbonate of lime,	31·94
Insoluble siliceous matter,	68·06
	100·00

It appears from these analyses—1. That the cornstones of Innerkip are carbonates of lime and not dolomites.

2. That in this locality igneous action has converted a carbonate of lime into a bisilicate. Now, as it appears from analyses 1 and 2 that there is but a trace of siliceous matter in the limestone, the origin of the silica must be sought in the igneous rock; in fact, a transference of a portion of its silica must have taken place when the basalt was in a state of fusion. Such transfer, indeed, could readily take place under the influence of chemical attractions, when the rocks were in a state of even imperfect fusion. On No. 4, or the table spar, Dr R. D. Thomson remarks:—

"It is an interesting fact in connection with this mineral, that it gives a yellow colour before the blowpipe when moistened with muriatic acid; and yields crystals of common salt when treated with the same acid. It is thus associated with Wollastonite, or soda table spar—a mineral occurring in the Bishopton tunnel, and in the Kilpatrick hills." There is, however, this difference between the two cases, that in the localities last mentioned the mineral has no direct connection with limestone, although calcareous spar occur abundantly; whereas, at Innerkip, it has obviously originated in a change induced upon common limestone.

V.—THE OLD SLATES.

151. These rocks are portions of the great bands of sedimentary strata which traverse Scotland from sea to sea, in a direction parallel to the principal axis of the Grampians, and to the great Caledonian valley. They first strike the shores of the Clyde a little to the east of Kilcreggan pier in Portkill bay, where the junction with the old red sandstone is concealed, as already mentioned. They deviate very little from their ordinary type, and it is therefore unnecessary to enter into any lengthened descriptions. The usual varieties, depending upon the varying proportions of the constituent minerals, occur abundantly, but in no definite order, and without much continuity. Thus the clay slate series exhibits beds of flinty slate often approaching to quartz rock, of highly bituminous slate, and of coarse-grained, compact, thick-bedded slate, mixed up irregularly with the commoner kinds, such as coarse and fine roofing slates, and a semi-crystalline silky slate, passing into chlorite or talcose schist. All these varieties are well seen on the road-side between Roseneath and Kilcreggan, and on the shore between the latter place and North Ailey. On the same coast thick cotemporaneous beds and also veins of quartz occur, in the cavities of which rock-crystal is often met with. Roofing slate of good quality is obtained from several quarries; but neither in this rock nor in the mica slate which underlies it, have any indications of

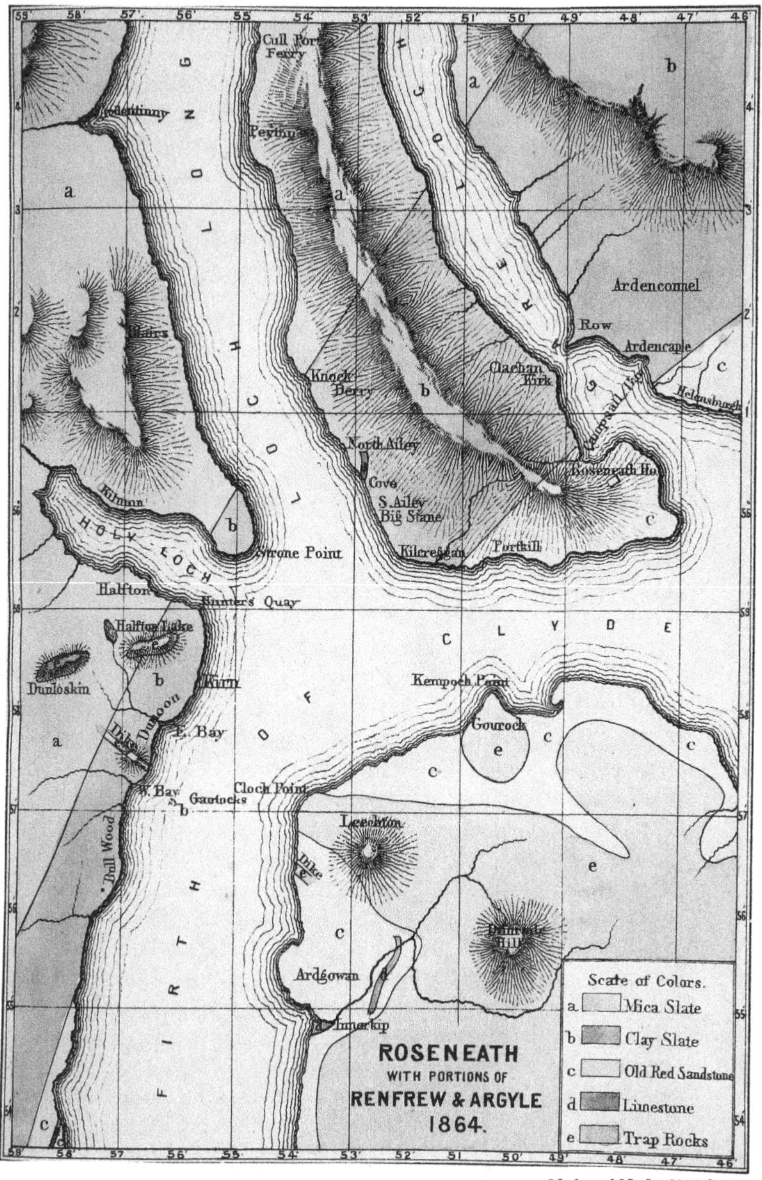

Maclure & Macdonald, lith, Glasgow

metallic ores been noticed. Iron pyrites occurs in the slate rocks in many places. Beds of quartz containing rock-crystal occur frequently in the mica slate, and are well seen on the road-side to the west of the village of Garelochhead. To the east of Tom-na-hary hill crystals of schorl are found in a variety of mica slate, containing very little quartz.

152. On a geological map these two slate rocks would be marked as separated by a definite boundary; but in nature no such distinction exists; the transition is in fact so gradual, that it is impossible to say where the micaceous series terminates, and the argillaceous commences. Towards its outer boundary the mica slate begins to assume the character of chlorite schist, and to contain occasional beds of fine roofing slate, the true mica slate still constituting the greater part of the mass. Farther out, the argillaceous and chlorite slates begin to prevail, so that the micaceous beds may be said to be subordinate to them; and thus through oft-repeated alternations, we at last reach the true clay slate series. We can conceive, therefore, of a certain middle line, along which the strata partake equally of both characters; it is this imaginary line which in such a case may fairly represent the boundary. These remarks are equally applicable to the other slate rocks of the district, and indeed to all the rocks of this class in the west of Scotland. In order to explain the mode in which this gradual loss of a marked character, and assumption of one considerably different was brought about, it is necessary to remember that the slate beds were originally deposited from the sea, layer over layer, in a state of silt or fine mud, and afterwards exposed to great heat, combined with pressure. A slight change in the nature of the sediment, or in the amount of heat or pressure, would be sufficient to produce the want of uniformity, and the variations from a definite type, which we now observe. The arrangement usually is, upper green or chlorite slate, middle dark-coloured clay slate, and inferior mica slate, in some tracts passing into gneiss, and extending inward to the Grampian granite. The mica slate has a

remarkable structural arrangement which distinguishes it from the two others; its crystalline laminæ are simple foliations, and form only one set of divisional planes; whereas in the other slates the bedding is traversed by a second set of planes, those of slaty cleavage, perpendicular to the bedding or stratification. The general direction of the beds is about north-east and south-west, the dip being to the south-east, at angles varying from 40° to 70°—but there are many local exceptions.

The great upward movement attendant on the elevation of the mica slate mass of Benlomond, has thrown the two other slates several miles southward of the line of bearing which they observe to the eastward of the mountain; and in like manner in other tracts along the Highland frontier, horizontal displacements on a great scale appear to have affected the two upper slate bands, so that the middle or dark slate is often seen in direct continuation of the upper chloritic band, or *vice versa.* And thus the appearance of frequent alternation is produced when these bands are traced from point to point on the line of bearing.

Plutonic and Metamorphic Rocks.

153. The slates above described contain subordinate beds and veins, which possess considerable interest.

To the right of the summit level of the road between Garelochhead and Portincaple Ferry, a remarkable ridge projects from the smooth outline of the hill-side, and trends in a straight line towards the base of the mountains. It consists of a highly felspathic rock erupted through the slate. The base is a very compact mixture of quartz and felspar, in which crystals of felspar and mica are imbedded, the latter ingredient being constantly present, the former often wanting. The prevailing colour is yellowish-red, given by the felspar in the base. From the prevalence of felspar, and the mode in which the crystals are disseminated, the rock must be called, according to the present views, a felspar porphyry; though the constant presence of mica as a constituent, and the compound character of the base, seem rather to require that it should be considered a granite.

At the sides of the vein next the slate the rock is of a more homogeneous character, resembling a compact claystone or flinty slate, but still enclosing crystals of mica. The laminæ of the slate in contact with the sides of the vein have a very close resemblance to this variety; being, in fact, a yellowish-gray, fine-grained, flinty slate, with occasional spangles of mica. It is thus difficult to determine the exact boundary line between the slate and the vein. This assimilation of mineral character has obviously been induced by the cooling of the masses from a state of fusion, at nearly the same rate, but more rapidly than the inner portions of the vein. The breadth of the vein is various; in some places no more than twelve or fifteen yards, in others as much as twenty-five, and even thirty, and perhaps considerably more, as it cannot in many places be exactly measured. It rises above the surface of the hill-side from 15 to 40 feet, and with several undulations runs in a general N.N.E. and S.S.W. direction.

In the northern part of its course it is interposed as a bed between the strata of slate. To the south of the high road, on the side of a little stream, it is seen intersecting the beds at a small angle. A little farther south the ground rises, and the vein does not appear upon the surface; but on the lowering of the ground still farther south, it again emerges from beneath the slate and occupies the surface for some distance. A similar overlapping occurs farther north, near the high road.

The appearance of this plutonic rock, now as a bed interposed among the strata, and again as a vein intersecting them, and the undulating course which it pursues, point out its posterior origin and the nature of the resisting force. This is plainly to be found in the peculiar undulations and twisted forms which everywhere characterise the mica slate, indicating the powerful compressing forces which acted upon it while yet plastic under the influence of heat.

Near the summit of the ridge, and at a curvature in the

vein, the outer or salient angle is intersected by a dike of greenstone and basalt, in such a manner that a portion of the felspathic rock is isolated between the whin dike and the mica slate on the west side, and the continuation of the vein lies on the same side of the dike as before the intersection. The annexed sketch, which is a *ground plan*, shews the mode of this intersection, one of the most singular we have ever met with.

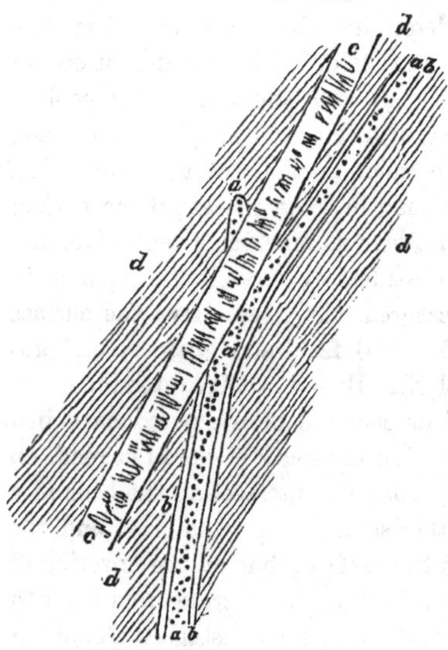

(*a a*) *Vein of felspathic rock.*
(*b b*) *Altered schist.*
(*c c*) *Dike of greenstone and basalt.*
(*d d*) *Ordinary schist.*

The dike is very distinctly traceable for several hundred yards towards the north-east, the surface occupied by it rising into conical hummocks. It is then lost in marshy ground for a short distance, but is again continued towards the mountains. In the other direction, its course was satisfactorily made out as far as Portincaple Ferry, where it is well seen; and a dike of the same width and bearing, which seems to be a prolongation of this dike, was met with near the top of the mountain on the east side of Loch Eck, traversing mica slate, and altering it considerably. The width is about 25 yards, and it bears a point S. of W. It is in many places inclined at the same angle as the slate, among the beds of which it seems to have insinuated itself in a serpentine course. The mica slate in some places is slightly changed by the contact, being rendered harder and more massive: the lamination is partially destroyed, and the rock is banded, parallel to the sides of the dike. In one place pieces of the slate are seen enclosed in the dike, and slightly altered. Portions of the wedge-shaped mass of slate (*d*), between the two veins, are entangled in the basaltic dike, and altered in the same manner. The circumstances of this case clearly prove the posterior origin of the basalt.

At the edges the dike consists of blue slaty basalt, but the greater part of the mass is a coarse-grained greenstone, which at several points exhibits in great perfection that peculiar structural arrangement in concretionary spheroids which is the most frequent and characteristic form assumed by the trap rocks, and of which the columnar is but the result, when under favourable conditions they parted more slowly with their heat of fluidity. The best marked of these is in a cliff about 60 feet high, overhanging the marshy ground above mentioned, where the dike has a considerable underlie, the slate being in contact on both sides. It is here divided into *columns of spheroids* perpendicular to the sides of the dike, and separated from one another by imperfect joints. Sometimes each joint is composed of a single spheroid; one was noticed measuring fifteen inches by ten; in other cases numerous small, closely-packed spheroids make up a joint. Instead of a distinct separation, as in basaltic pillars, the columns are connected by narrow seams of decomposed greenstone. The columnar structure is here seen in the act of development. If the heat had parted more gradually, a façade of pillars would have been the result. This spot affords an excellent illustration of the remarkable experiments of Mr Gregory Watt on fused basalt (*Phil. Trans.*, 1804; see Art. 52).

It will appear from the foregoing statements, that the small area we have been describing is one of considerable interest, exhibiting, as it does, the rare association of many species of erupted rocks in connection with the primary strata; and affording illustration of some curious questions in theoretical geology.

Limestone.

154. On the shore of Loch-Long, north of Kilcreggan, and about 250 yards south of the landing pier at Cove, a bed of limestone is interstratified with the clay slate. It has been originally six or seven yards wide, and has extended eastwards across the low ground between the shore and the cliff, into the cliff itself, and probably much farther inland; but it cannot be satisfactorily traced. The part next the shore has been almost entirely removed by quarrying; but from portions which are found among the slate—as shewn in the annexed sketch, which is a *vertical section*—there can be no doubt of the true position of the bed.

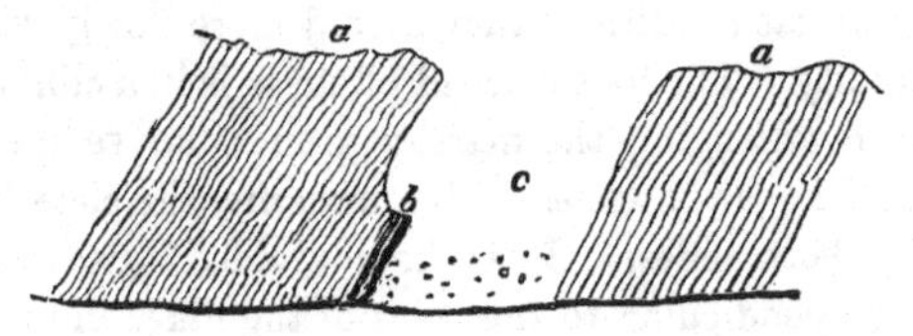

(a a) Inclined strata of clay slate; (b) bed of limestone; (c) bay, with accumulations of shingle.

The limestone is impure, from intermixture with slaty laminæ; the prevailing colour is bluish-gray; it contains much calcareous spar, and, like the slate, is destitute of fossils.

In the New Statistical Account of the parish of Row adjoining, beds of limestone are stated to occur in the slate rocks of Glenfruin; these are most probably similar to the bed now mentioned.

In a paper by Mr Daniel Sharpe, F.G.S.,* many limestone beds in various tracts, which Dr MacCulloch and other observers have described as belonging to the primary strata, are referred to the old red system; and he expresses a decided opinion with regard to all of them, denying the occurrence of such beds in association with at least the clay slate system. The cases here given clearly disprove this view; and others in the Highlands are known to us.

* Jour. of Geol. Soc., 1852.

Slates of the Cowal Coast.

155. This coast consists of slates of the chloritic and argillaceous series, passing westwards into mica slate. The slates near Dunoon are associated with rocks of igneous origin, to whose effect upon the slates, and their own peculiar forms, much of the picturesque beauty of this favourite watering-place is due. Thus the ridge lying between the coast and the valley of Hafton lake, owes its elevation and bold outline to an outburst of igneous rocks, which have induced a very decided change upon the slate along the planes of contact. It consists of crystalline greenstone, of a different type from that of the dikes common on the coast, the structure being slaty and the hornblende in excess. It is from 60 to 100 yards wide, and ranges from near Hunter's Quay, across the highest part of the ridge, transversely to its length, appearing along the summit in a series of conical hummocks, with deep hollows between; and thus presenting a bold, picturesque outline when viewed from the low grounds in the neighbourhood. It is interrupted by the Hafton valley, but is resumed on its western side, and attains its greatest altitude in Dunloskin hill, which rises prominently above the surrounding slopes, strikingly relieving their monotonous outline. Westwards, for about half-a-mile, it is seen in other rocky eminences, but its farther extension in this direction was not traced. The ridge is not seen intersecting the coast, which is everywhere occupied by the slate rocks; so that it seems to terminate before reaching the shore. Owing to the metamorphic character which has been impressed upon the adjoining slaty beds, it is difficult to determine the precise limits of the plutonic rock; near the contact the slate breaks under the hammer into very compact four-sided prisms.

In a similar manner, the high ground dividing the East and West bays, and projecting beyond the general line of coast, has acquired its strikingly picturesque aspect from a great dike of basalt which traverses it. The Castle hill consists of this dike, and of slate borne up with it, and adhering to it. By contact with the dike, the slaty structure is effaced; the rock has been fused and reconsolidated into a compact flinty slate, closely resembling basalt; crystals are developed along the boundary, and bands of different colours are disposed parallel to the sides of the dike. The width is about 100 feet, and the bearing W.N.W. The Gantock rocks are exactly in the line of bearing, but were found to consist of very hard slate. On the opposite coast, however, near Ardgowan, a dike of the same width and direction occurs, which may be the continuation.

Near Innellan the slates are covered by patches of old red sandstone and its associated cornstone, as already mentioned.

VI.—THE CLYDE ISLANDS.

Bute.

A pretty full and generally accurate account of the geology of Bute has been given by Dr MacCulloch in his work on the Western Isles, published in 1819. For thirty years after, no observations, so far as we can learn, were put on record respecting it, except some notices of its remarkable coast terrace and raised shelly deposits by Mr Smith of Jordanhill. The greater variety of the strata in Arran, and the bearing of the phenomena there exhibited upon questions in theoretical geology actively discussed at the time, drew attention entirely to that island, and Bute, in common with other parts of the west of Scotland, was overlooked. Yet it has many points of great interest; strata occur here to be met with nowhere else in Scotland; and the effects of its trap dikes upon the adjoining strata are extremely curious, indeed of a unique character.

156. Three deep depressions or valleys traverse Bute perpendicularly to its longer axis, dividing the island into four portions, and marking the boundaries of as many distinct geological formations. They terminate on either side in bays or indentations of the land, formed here as in most other cases at the points of least resistance, the junctions, namely, of dissimilar strata. Those on the east side are the well-known

sheltered bays of Kames, Rothesay, and Kilchattan. The low tracts in question shew no rock *in situ*, but are filled with shingle and alluvial deposits concealing the junctions, strata of peat, and occasional shell beds. The elevation above the sea level nowhere exceeds 30 feet; and as this is also very nearly the height of the terrace already referred to as encircling the island, it appears that when the sea stood at that ancient level, Bute consisted of four islands, separated by narrow channels.

The various strata exhibited in Bute are the terminal portions of those great bands of rock, sedimentary and igneous, which extend across the country from sea to sea, as already noticed. Mica slate occupies the northern portion between the Kyles on the north, and Kames and Ettrick bays on the south. The rock has its usual character and aspect, and rises

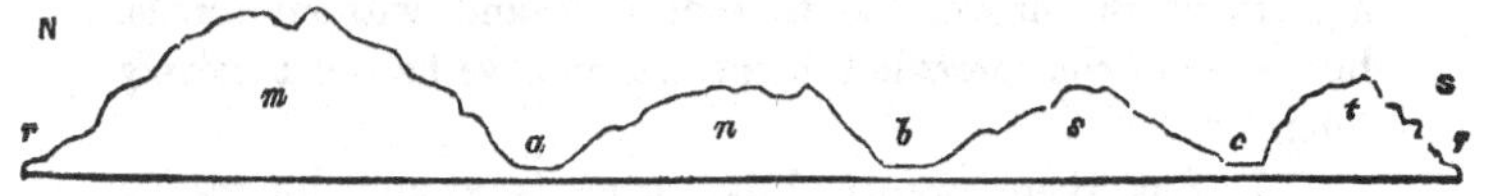

(a) Kames bay; (b) Rothesay; (c) Kilchattan; (m) mica slate; (n) clay and chlorite slates; (s) red sandstone; (t) trap; (r) the terrace.

into hills of nearly 1,000 feet elevation. The district south of this, bounded by the Rothesay valley, consists of the two upper slates, the common clay and chloritic. Subordinate to these are great beds of quartz rock, the most considerable of which forms the high ridge called Barone hill, with a picturesque old ruin overlooking Rothesay on the west. There are also copper veins in this slate, westwards from Kames bay. The portion extending from Rothesay valley to Kilchattan is occupied by red sandstone; and finally the southern portion, with a substratum of red sandstone, consists mainly of various rocks of the trap family, erupted through and overlying the sandstone. The accompanying outline of the island shews the relation of these strata to the valleys or depressions, which are obviously a part of that system of parallel fractures ranging N.E. and S.W. on both sides of the Grampians, and probably due to the upheaval of this chain, and the later igneous eruptions already noted.

A description of these rocks would be useless, as it would merely be a repetition of facts contained in all elementary works. The great body of sandstone obviously belongs to the old red system; and as well from its position in immediate sequence to the old strata, as from its general mineral character, seems to form the lower portion of that system; but hitherto no fossils have been detected in it. Like the corresponding strata on the mainland, at the base of Benlomond, Innerkip and Innellan, it contains subordinate beds of cornstone. These are seen in several places along the shore south of Bogany point; but the beds are thin, and generally contain much siliceous matter, so that the limestone is of no economical value. The Kilchattan beds are of the same age; they are subordinate to sandstone and without fossils, but being of considerable thickness, they have been extensively quarried.

157. But there are in Bute sandstones and limestones newer than these, which have been till very recently quite overlooked by geologists. There occurs, in fact, hanging on to the flanks of the old red sandstone at Ascog, a small coal formation, identified by means of its fossils with the lower marine series, which may yet turn out to be of economic value. It is connected with an isolated, overlying mass of trap appearing on the shore, and occupying the cliffs near Ascog mill. On the north side of the promontory, south of the mill, several thin courses of nodular limestone traverse beds of brown-coloured, crumbing shale subordinate to sandstone. The shale is of considerable thickness, and rises into banks above the road. The south side of the promontory presents the following section:—

The lowest bed (*a*) is a fine-grained, bluish-gray, nodular limestone. Over it is a bed of black bituminous shale (*b*), containing veins of coal about a quarter of an inch thick; and upon these rests a bed of concretionary limestone (*c*), the base or paste being a dark-coloured limestone, and the concretions rounded lumps of the same rock, often of considerable size. The upper part of the cliff is occupied by trap in various prismatic forms. The base of the concretionary limestone is so much altered by the contact of the trap, that the two rocks can only be distinguished by the action of a strong acid. A like change is produced upon the imbedded lumps in the upper part of the bed. The limestone, shale, and coal

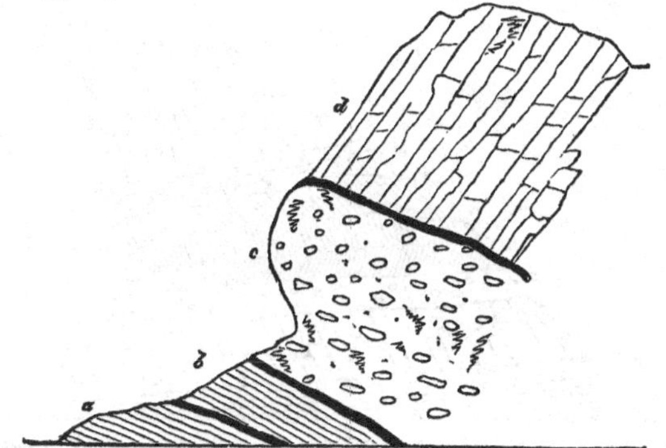

(a) Limestone; (b) shale with thin coal seams; (c) limestone breccia; (d) trap.

seams extend under high-water mark, and when the tide is very low considerable pieces of coal are often dug out from beneath the sand and mud covering the tide-way. Several sinkings have been made here for the purpose of discovering workable coal seams, but without success.

158. The trap above referred to is a projection from the principal mass above, which occupies a considerable area inland towards Ascog lake, and is upwards of 100 feet thick. It is seen in the summit of the cliffs on either side, the line of junction ascending rapidly as it retires from the shore. These trap rocks at Ascog derive their chief interest from being the repository of beds of lignite—a substance so rare in Scotland, that no well-marked beds occur on the mainland, and but few in the other islands; and these in situations very difficult of access.

The principal bed is situated in the face of the cliffs above the road, a little to the south of Ascog mill, as shewn in the annexed section giving the various beds:—

The lowest bed resting on the sandstone is a small-grained, rudely columnar greenstone; the junction is, however, concealed. Over this is a trap-tuff with a base of greenstone, and imbedded spherical lumps of the same substance. This is followed by a bed of red ochre of coarse texture, traversed by numerous black

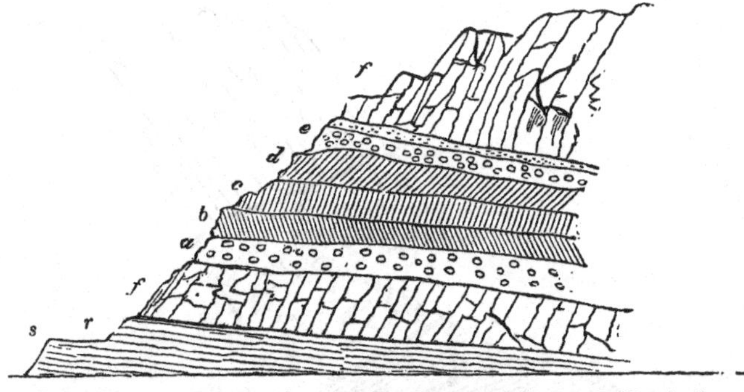

(s) Sandstone; (r) terrace and road; (ff) greenstone; (a) trap-tuff; (b) red ochre; (c) lignite bed; (d) pisolitic ochre; (e) porphyritic amygdaloid, the upper portion much altered.

iron seams, which have doubtless been produced from a change in the oxidation of the component iron. Over this is the lignite bed: it is three feet thick, and consists of hard stony coal, interstratified with a yellowish-white shale, both being much intermixed with pyrites. The coal has been so much altered throughout its whole thickness by the contact of the trap rock, that Mr Rose of Edinburgh, to whose examination the best specimens that could be selected were submitted, in order that he might determine the species of wood, but without any note of the geological situation of the coal, was "unable to obtain a slice, in consequence of the structure being altered by the contact of a whin dike." The coal has been worked to some extent by driving an adit inwards on the line of the dip, which is about 20° to the westward; but the workings have been for some time abandoned, and the inner and lower portions are now full of water.

The floor of the coal has been already described. The roof is a peculiar rock. It consists of a base or paste of an ochreous steatite, with imbedded round pieces of the same substance, and may hence be called a pisolitic ochre; it is three and a-half yards thick. The bed above this is of the same character; but the base feels less unctuous, and with the imbedded steatite it contains also imbedded calcareous spar. The base effervesces briskly with an acid; and hence we may call the rock a calcareous amygdaloid. The upper portion of this bed, to the thickness of a few inches only, is very hard, and has a semi-vitreous appearance, and thus closely resembles a porphyry. In common with the trap above, and indeed all the beds in this locality, it contains much disseminated iron. The rest of the cliff is occupied by greenstone, similar to the lower bed in contact with the sandstone.

Another bed of lignite occurs on the opposite, or north-west side of the trap district, overlooking Ascog lake. The coal dips to the interior of the area, that is, nearly south. It is of about the same thickness, and is accompanied by beds of steatite and red ochre, very similar to those above described; but the nature of the ground is such that a complete section cannot be had, and the precise number, therefore, and order of the beds, cannot be exactly stated. The association, however, of the lignite with ochres and steatites here also is sufficiently distinct, and it is even probable that these beds are persistent throughout the whole of this district.

In the basaltic district of the north-east of Ireland, lignites occur in the middle and upper parts, associated with variegated ochre and common trap. There is, as we have seen, a like association in Bute. The interesting leaf-beds and associated lignites, overlaid by trap in Mull, discovered and described by the Duke of Argyle (*Jour. Geol. Soc.*, 1851), seem to be a deposit precisely analogous to the Antrim and Bute lignites; and from casual observations of Dr MacCulloch it would appear that such beds occur also in Skye and others of the Western Isles. Similar conditions thus appear to have prevailed over a very wide area, the successive eruptions of igneous matter over the sea bottom were very similar, and there were like periods of repose, during which the productions of the adjoining land were swept down to be buried under the next flow of submarine lava.

The wood which has supplied the Antrim and Mull lignites, has been ascertained to be coniferous; but the species have not been determined. Mr Robert Brown and Professor Lindley referred the former with some hesitation to one of two species—the common fir (*Pinus Abies*) or the Weymouth pine (*P. Strobus*); and Professor Edward Forbes, who reported upon the Mull leaves, at the request of the Duke of Argyle, found that they belonged to a *taxus*, a *platanus*, and several species of *rhamnites*, upon which he did not venture to pronounce positively (*Jour. Geol. Soc.*, Vol. vii., p. 103, 1851). The lignites which occur in Mull in a bed under basalt, apparently in direct continuation of the leaf-beds, have been recently submitted by Professors Harkness and Blyth to a microscopical and chemical examination, and their structure compared with that of the Causeway lignites (*Ed. Phil. Jour.*, N. S., Vol. iv., p. 304, 1856). The results prove that the variations from the ordinary coniferous structure are the effects of the great pressure to which they were subjected by the accumulation over them of thick beds of igneous matter. We have failed, as already remarked, in determining the nature of the Bute lignite; but it has quite the appearance of being coniferous; and the beds associated with it being exactly the same as accompany the Antrim lignites, there is every probability that both are of the same age. Now, the basaltic series of the north-east of Ireland, as it overlies the chalk formation, clearly belongs to the tertiary era, and was long ago recognised as of this age. But many cases occur in which the same basaltic flow, which alters the chalk to the state of a saccharine marble, spreads out beyond the limits of the chalk, and overlies and alters the new red sandstone and coal measures. Such overlying masses must also belong to the tertiary period. It is therefore not improbable that the basalts forming the lofty cliffs on the west coast of Mull, superimposed upon the lias and oolites, may also be tertiary. The Duke of Argyle is inclined to refer these to an older date; while his Grace, in the able and interesting paper above mentioned, clearly shews that the Ardtun leaf-beds and lignites are of tertiary age. It seems to us, however, to be premature to place these and the Antrim lignites in the miocene group, as his Grace and Sir Charles Lyell have done, until greater certainty is attained with regard to the species to which the leaves and lignites are to be referred.—It is not then improbable, that our Bute trap formation, with its ochres and lignites, may also be of tertiary age, though overlying the older secondary strata.

The chemical composition of ochre is almost exactly the same as that of basalt; 100 parts consist of—silex 56·40, alumina 3·46, per-ox. iron 24·14, carb. lime 0·90, water 15·10. It is, in fact, a decomposed basalt, or volcanic ash, partially re-fused and reconsolidated by the succeeding flow of igneous matter.

159. The limestone of Kilchattan bay is a cornstone, subordinate to the old red sandstone; the strata of the two rocks are conformable, and the dip nearly south at a moderate angle. At the summit of the ridge, near the picturesque ruins of the ancient castle of Kelspoke, the sandstone over the limestone is seen dipping under the trap which bounds the rugged terraced ridges descending towards Garrochhead on the south coast. These ridges have the same inclination southwards as the underlying sandstone strata, and present a succession of bold fronts towards the north. The sandstone appears to have had its present inclination when the submarine lava streams, of which these ridges consist, were poured out over it; the scarped fronts were no doubt formed by the action of currents when the land was rising. The arrangement of the strata at Kil-

chattan is shewn in the annexed cut.* No fossils were found in the limestone.

160. The dikes of Bute are composed of greenstone, or basalt, and are very numerous, especially on the east coast. They traverse the strata in various directions, and in some cases can be traced for several miles continuously, preserving nearly the same width and direction throughout.† Two or

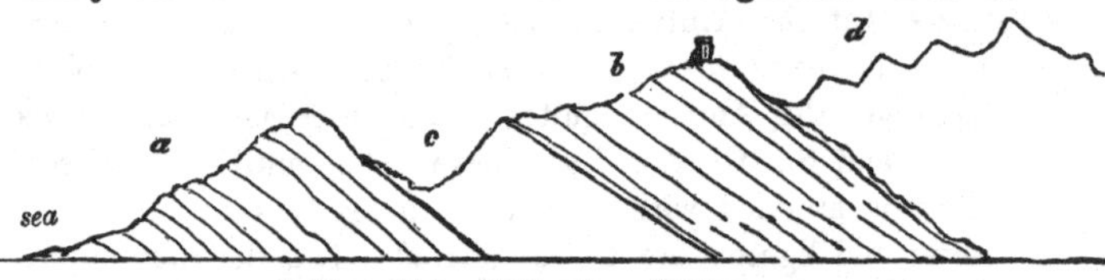

(*a*, *b*) *sandstone;* (*c*) *limestone;* (*d*) *trap.*

more are sometimes seen to meet and coalesce for some distance, and again to separate;—a narrow dike branches off into several filaments, which unite again—portions of the rock traversed are often found entangled in the dike; and these, as well as the contiguous strata, present the usual metamorphic effects recognised as due to igneous action, besides others of a peculiar and exceptional character, to which we shall now allude.

The Kilchattan limestone is altered in a remarkable manner by a large dike, crossing it nearly in the direction of the dip, and the effects are well seen at the eastern side of the quarry.

* This and the other cuts illustrative of Bute are copied, by permission, from the *Journal of the Glasgow Philosophical Society* for 1848.

† A dike seventy feet wide emerges from the sea at the mineral well near Bogany point, and ranging nearly west is seen in Huntly place, where it has been largely quarried; interrupted by the bay, it rises again, is conspicuous across the high grounds west of the town, and crossing the island, enters the sea at Ettrick bay. Here another dike, two or three times the width, enters near it, crossing from Ascog, and visible in several eminences in the interior. The Rothesay mineral water, which has acquired some reputation for efficacy in rheumatic, cutaneous, and glandular complaints, rises in the former of these two dikes at Bogany point. The gallon of 277·274 cub. inches contains, according to the analysis of Dr Thomson,—com. salt, 1860·78 grains; sulph. lime, 125·20; sulph. soda, 129·77; mur. magnesia, 32·80; silica, 14·39; sulphuretted hydrogen, 17·4 cub. inches. Both dikes and this mineral water are noticed at some length in *Wilson's Guide to Rothesay and the Island of Bute*, an excellent and neatly illustrated little work.

Along the plane of contact with the dike the limestone is altered to the state of a granular saccharine marble, which on the application of a slight pressure crumbles into a fine powder. This is succeeded by a hard crystalline marble, the crystals appearing in distinct flakes. Between this and the first change, which is one of simple induration, there are many gradations. Similar effects are common at the contact of limestone with plutonic rocks; in some localities they are accompanied by other singular changes of a chemical nature. Magnesia, and sometimes silica and alumina, are introduced into the composition of the limestone, so that simple carbonate of lime becomes a double carbonate of lime and magnesia. The source whence this magnesia has been derived has occasioned much difference of opinion among geologists. Some imagine that it has been transferred from the plutonic rock to the limestone; while others hold that, as fractures and dislocations of the earth's crust accompanied the eruption of these plutonic rocks, gaseous exhalations might find their way from beneath, and introduce carbonate of magnesia and other substances into rocks near the surface. In confirmation of this view, Mr Phillips has shewn in his *Geology of Yorkshire*, that "common limestone is dolomitised by the sides of faults and mineral veins far away from igneous rocks of any kind;" and some distinguished chemists have expressed their belief that carbonate of magnesia may be sublimed by the action of great heat. (*Rep. Brit. Assoc.* for 1835, Trans. Sect., p. 51; Phillips's *Geology*, in *Cab. Cyclop.*, Vol. ii., p. 98.) Much doubt, however, still hangs about this subject. Cases occur in which magnesia has been introduced, although the limestone could not have been subject to such a pressure as would confine its carbonic acid when the rock was softened by heat.

In order to elucidate, if possible, this obscure subject, two specimens of the rock were submitted to Mr John Macadam, lecturer on chemistry, now of the Melbourne Philosophical Institution, for examination with reference to the presence or absence of magnesia. Specimen No. 1 is the saccharine marble from contact with the dike; No. 2 is the unaltered limestone —both average specimens:—

"In specimen No. 1, carbonate of magnesia constitutes about 2½ per cent. of the whole mass. Its other and principal ingredients are carbonic acid and lime, silica, and traces of oxide of iron and alumina.

"In specimen No. 2 magnesia abounds, the amount present being equivalent to 33·72 per cent. of carbonate of magnesia. The other constituents are similar to those in No. 1. From the large proportions of carbonate of lime and carbonate of magnesia, No. 2 would appear to be a species of dolomite. It may be noticed that the physical characters of No. 2 are very different from those of No. 1; the former is difficult to pulverise, the latter is extremely susceptible of division.

"The action of strong hydrochloric acid on both specimens causes a portion of gelatinous silica to appear, shewing the presence of a silicate, which may be that of magnesia, since the quantity of gelatinous silica is about sufficient to combine with the 1·28 per cent. of caustic magnesia existing in the specimen No. 1. There is a less quantity of this gelatinous silica in No. 2. The greater portion, however, of the silica present in both specimens remains undissolved in the gritty or pulverulent condition, and is hence in a state of mere mechanical mixture with the other constituents of the limestone. It would require a minute quantitative analysis to determine whether the 1·28 per cent. of magnesia exists as a carbonate or silicate, or partly as both."

The phenomena are thus of a contrary character to what is usually found; the unaltered rock is a dolomite, and contains nearly 34 per cent. of carbonate of magnesia, while the altered rock contains less than 3 per cent. What has become of the constituent magnesia? Has it been driven off by the heat to which the limestone was exposed? Most chemists are unwilling to admit that this is possible; and it may reasonably be objected, that if the limestone had been exposed to so high a temperature as to vaporise its magnesia, the silica would not be mechanically present, but would have entered into chemical combination with the lime or magnesia, and have formed a silicate.

That whin dikes have sometimes been the means of producing such a combination has been shewn by an eminent chemist. Dr Apjohn found the white chalk of Antrim, altered to the state of a saccharine marble by whin dikes, to be a trisilicate of lime, "very analogous in its composition to olivine. We are thus enabled to understand why olivine should be so very frequently found in trap rocks, and to refer its origin to the contact of silex at a high temperature with an excess of the basic oxides; and we have in some degree a demonstration that the dolomites which contain siliceous sand could not have been exposed at any time to a heat sufficiently high to account for the introduction into them of magnesia in the vaporous state; for by such a heat a silicate of lime or magnesia, or of both, would have been produced" (*Jour. Geol. Soc. Dub.*, Vol. i., p. 376).

The presence of these silicates in both our specimens is shewn by the appearance of the gelatinous silica; yet a greater quantity of silica is present mechanically, which, as already stated, seems inconsistent with the exposure of the rock to intense heat; unless, indeed, we could suppose that the silica has been introduced by infiltration, or the magnesia removed by the solvent power of free carbonic acid at a period subsequent to the consolidation of the dike from a state of igneous fusion.

Careful *quantitative* analyses of the limestones were made by the late Dr Robert D. Thomson. It is hoped that these will afford definite terms of comparison with other analyses, such as those of Dr Apjohn, already referred to; and that their publication may lead to the formation of clearer views respecting an obscure question in theoretical geology.

The analyses are as follows:—

Specimen No. 1 is the saccharine marble from contact with the dike at Kilchattan—in the highest state of alteration.

No. 2 is the hard crystalline marble, having the crystals in distinct flakes, more remote and less altered than No 1.

No. 3 is the unaltered limestone from the middle of the quarry, remote from the dike—an average specimen.

No. 4 is the altered limestone from contact with the overlying trap at Ascog mill; it is an impure, dark-coloured rock, of an earthy aspect, and very like the trap which rests upon it.

No. 1.—Spec. grav. 2·710.

Silica, Alumina,	6·91	5·16	5·70
Protoxide of iron,	1·68	1·50	1·28
Carbonate of lime,	90·65	—	91·08
Carbonate of magnesia,	1·00	—	1·17
	100·24		99.23

No. 2.—Spec. grav. 2·570.

		I.	II.
Silica, Alumina,	1·94	0·28	0·28
Protoxide of iron,	0·52	—	0·56
Carbonate of lime,	96·48	98·76	96·58
Carbonate of magnesia,	1·23	—	2·24
	100·17		99·66

No. 3.—Spec. grav. 2·679.

	I.	II.	
Silica, Alumina,	9·70	—	9·08
Protoxide of iron,	1·12	—	1·12
Carbonate of lime,	67·42	72·12	67·00
Carbonate of magnesia,	17·31	—	18·06
Water, coaly matter, and carbonic acid,	4·45	—	4·74
	100·00		100·00

No. 4.

Silica, Alumina,	64·21	64·46
Protoxide of iron,	6·42	6·60
Carbonate of lime,	24·00	21·20
Carbonate of magnesia,	4·62	2·85
Water and carbonic acid,	1·75	4·89
	100·00	100·00

The silica present is in a state of mechanical mixture.

These analyses are confirmatory of the main points of the views already stated, and seem clearly to establish the new and remarkable fact, that by the igneous action in these instances the magnesia has been driven off from the limestone. The unaltered rock is a dolomite containing nearly 70 per cent. of carbonate of lime, and nearly 20 per cent. of carbonate of magnesia; while the altered rock contains but from 1 to 2 per cent. of the latter ingredient. To what cause are we to assign the changes that have taken place? Has the magnesia been sublimed by heat? or has it been withdrawn by the solvent power of free carbonic acid? On the nature of these and the other chemical changes that have been induced, it is difficult to express an opinion; and from such limited premises it would be unphilosophical to draw any general conclusions. The subject is one, however, of great interest both to the geologist and chemist, as the facts are directly opposed to the received views; and as no instance of similar changes on dolomitic rocks has, so far as we are aware, ever been put on record.

161. A remarkable dike intersects the sandstone between

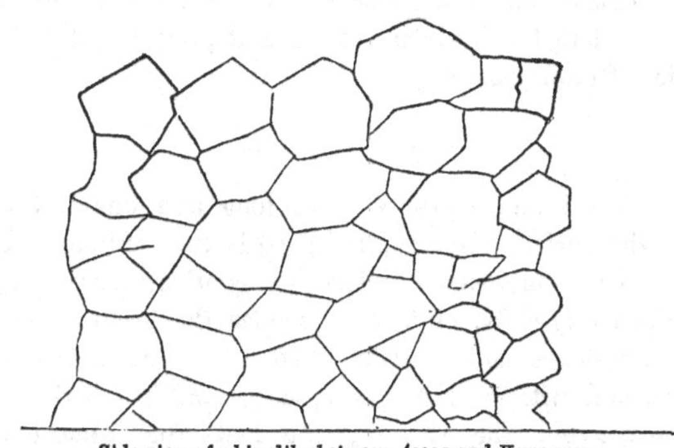

Side view of whin dike between Ascog and Kerrycroy.

Ascog and Kerrycroy, well worthy of attention on account of the striking illustration it affords of the mode of cooling of basaltic rocks from igneous fusion. It runs parallel to the shore for some distance, and then retires from it towards the S.W., and striking the inland cliff already mentioned, whose direction is here the same as that of the dike, it forms the perpendicular face of the cliff in front of the sandstone, rising like a wall to a height of twenty or thirty feet. The direction of the cliff soon changes, however, and the dike then enters the hill behind, and is lost. The sandstone having been completely worn away from the seaward face of the dike, a very large surface of the side is laid bare, and the structure is exhibited as in the annexed cut.

The dike is composed of greenstone, is about fifteen feet wide, and the prisms are mostly pentagons and hexagons. The cause of this peculiar structure has been already explained (Art. 52).

The Arctic shell-beds of Bute, and the evidences of glacial action, have been noticed already (Arts. 123, 126, 133); and further details need not be here entered on.

Inchmarnoch.

162. This small island lies on the western prolongation of the outer slate band which crosses Bute, and is noted for its quarries of roofing slate.

Great Cumbrae.

163. This island is three miles long and one and a-half broad; the surface rises gently towards the middle, and the height nowhere exceeds 500 feet; the southern shore presents some low rocky cliffs; on the other sides the ground descends by grassy slopes and sandstone ledges. Wanting trees and streams, and without lofty rocks, the island has little variety of scenery; but the views to be had from almost every part of it are of surpassing magnificence and beauty. Great Cumbrae corresponds in geological structure with the middle region of Bute, and consists of red sandstone,

having a slight northerly dip, traversed by numerous trap dikes, invaded by sheets of the same rock, but not overlaid by trap except in one spot on the west coast, where a small patch occurs. The sandstone is generally of a deep red colour; almost the only exception being found in the small islets forming the harbour of Millport, which consist of white and gray sandstone. Strata of conglomerate structure are occasionally met with, and the ridges on the west formerly shewed some thin limestone beds interstratified with the sandstone, but these have been worked out. No fossils have been met with; but there can be no doubt that the sandstone, like that of Bute and the mainland opposite, is a member of the Old Red series. The only varieties of trap occurring here are coarse and fine greenstone, and basalt; the fine greenstone has often a porphyritic structure, from the imbedded felspar crystals, thus forming the variety incorrectly termed trap porphyry. The chief interest of the island is in its whin dikes. These alter the strata remarkably: the sandstone has been fused by them, and reconsolidated into a substance closely resembling a dark quartz rock; simple induration is induced at a greater distance from the dike. Many of the dikes stand out boldly from the adjoining sandstone, which has been worn away on either side, the amount of wearing in the ancient tideway, compared with that in the present, affording a rude measure of the time during which the sea remained at the higher level. Attention was first called to these remarkable dikes by Mr Smith of Jordanhill. The best example is seen a little to the east of the entrance of Millport harbour, where a large dike rising from the sea level like a huge wall runs far up along the hill-side, raised as if by art above the surface of the fields. On the shore of the mainland opposite, another dike, having the same direction and apparently a prolongation of this, stands out in the same manner from the surface of the sandstone. The similarity of the two masses of rock and their former connection are pointed at in the legend that this strait was once spanned by an enormous bridge raised by the hands of mighty wizards, of which the only portions that remain are these two ancient abutments.*

Such projecting masses of trap are not confined, however, to the shore; they occur in the inland parts, where they are, of course, not so easily accounted for. We can hardly suppose them due to atmospheric causes, as there are no streams to carry off such worn materials; and they are most probably due to the action of the sea upon the sandstone during the progress of the last elevation of the land.

LITTLE CUMBRAE.

164. The Lesser Cumbrae is about two miles long and one mile broad, and in its bolder outlines strongly contrasts with the sister isle; its highest point has an elevation of about 800 feet. In geological structure and in altitude it corresponds with the southern division of Bute; and, like the Garrochhead district of that island, consists wholly of trap—sandstone only appearing in one spot on the east shore, near the old keep. This patch, however, is sufficient to shew that the foundation of the island is made of the same sandstone which forms the Great Cumbrae and the adjoining mainland. Piled up over the sandstone foundation in great successive sheets, these trap rocks give a terraced and ridgy structure to the island; and, rising south-west, present steep, abrupt cliffs towards the sea, while they decline in succession in a N.E. direction, or contrary to the inclination of the rocks of the Garrochhead. This is, no doubt, due to the original inclination of the surface over which the submarine lava streams were spread out; but that inclination is most probably due to a great fault in the sandstone, on which the sea, working more successfully along a fracture, opened this main entrance into the proper estuary of the Clyde.

There is a very interesting variety among the trap rocks here, and they often approach the form of perfect columns. They differ from the Garrochhead traps; and are much more closely related in mineral structure to those of the mainland.

* Quart. Jour. Agric. and Trans. High. Soc., No. xliii.; or Vol. ix., p. 430.

Mr Smith of Jordanhill has remarked, that in exact correspondence with the greater hardness of the rock, the terrace in front of the old sea-cliff is narrower than in Great Cumbrae; and he finds glacial striæ marked upon the surface of the terrace parallel to the cliff, and extending, still unworn, beneath the present level of the sea (*Newer Pliocene Geol.*, p. 144, and *Jour. Geol. Soc.*, 1862).

The small keep, on a peninsula on the eastern shore, is one of a range of watch-towers erected along the Clyde shores during the wars of the Edwards.

Pladda.

165. This islet, we have seen already, is attached to Arran by a whin dike, over most parts of which there is broken water at low tide. The island consists almost wholly of a dark-coloured trap rock, the sandstone foundation appearing only on the north-eastern shore.

Ailsa Craig.

166. This majestic rock rises steeply from the sea to the height of 1,097 feet. It is 13 miles S. of Pladda, and 10 miles W. of Girvan, and has an elliptic base 3,300 feet by 2,200. Its form is nearly that of a right cone; but it is somewhat more elongated from north to south than in the opposite direction. This is distinctly seen on climbing to the summit, which presents a flattened ridge in the former direction, and is comparatively narrow from east to west. In shelter from the prevailing winds, a high bank of shingle has accumulated in the course of ages against the originally steep face on the east side, and is now perfectly stable. From the top of this bank the rock rises in broken terraces, up which the ascent is not difficult; on other sides it is inaccessible. An old stone keep, with vaulted apartments—one of the line of watch-towers—is perched upon a terrace on this side, at about one-fifth of the height. With the exception of the shingle bank, the isle consists wholly of a reddish-coloured,

close-grained syenitic greenstone, quite unlike any other of the Clyde traps. On the Ayrshire coast, opposite to Ailsa, a similar syenite is said to occur amid strata of silurian age. The rock consists of red felspar and hornblende, and occasional grains of quartz, and is intermediate between a true syenitic or hornblendic granite and syenitic greenstone. It is thus of igneous origin and a member of the trap family, though not a basalt, as often stated. The rock, however, affects the columnar form, and the precipices on three sides exhibit magnificent ranges of pillars, which reach the height of fully 400 feet; but these are much less perfect than in basalt; the angles are less sharp, and the concavo-convex joints are wanting, the divisions being simple seams. The upper 700 feet consist of amorphous rock, but there is no perceptible difference between the mineral structure of this and the columnar part. The precipices are traversed by numerous basaltic dikes, which simply separate the rock, without inducing any change of structure upon it.—Mr William N. MacCartney, of Glasgow, has discovered on Ailsa in the present year the rare plant *Lavatera arborea,* or tree mallow. It grows high on the southern cliffs, and reaches the unusual length of 6 feet. He has carefully collected all the plants during two visits—our own was too hurried to admit of this—and finds 109 species of flowering plants, of which 14 are grasses. The cryptogamic flora is unusually abundant for a space so limited; the peculiar situation, but still more, perhaps, the terraced character of the sheltered side, favouring a prolific vegetation.

TABULAR ST

Shewing the valuable Seams of Coal and Ironstone in the Glasgow Coal Fields, wit

Depth in fathoms as at Glasgow.	SEAMS.—NEW RED SANDSTONE.	LOCALITIES WHERE THI
40	Palace Craig ironstone, ...	Palace Craig, Carnbroe, ...
48	Upper coal, ...	Glasgow, Stonelaw, Eastfield, Rutherglen, ...
60	Ell coal, ...	Glasgow, Stonelaw, Eastfield, Rutherglen, Bail
64	Pyotshaw and Main coals, ...	These seams are together at Airdrie, Coatbr separated a distance of from 5 to 30 feet Pyotshaw is a thin, unworkable coal in the
73	Humph coal, ...	Glasgow and Airdrie. A comparatively useless
77	Splint coal, or Lady Anne, ...	Glasgow, Airdrie, Baillieston, Wishaw, Mother nace coal of the district, and was for many in Lanarkshire, ...
78	Sour milk coal, ...	At Glasgow, Rutherglen, Stonelaw. The Wee c coal of Baillieston and the Shettleston dist
93	Mushet black band ironstone, ...	Airdrie, Coatbridge, Arden, Cleland. This is th found, however, out of the district immedi towards Drumpeller on the west, Rochsoles
95	Soft black band ironstone, ...	Worked at Airdrie, where it was found of inferi
104	Rough band ironstone, ...	Worked in the districts adjoining Cleland and
106	Virtue-Well coal, ...	At Airdrie, Coatbridge, Greenhill, Cleland, Coa coal of Slamannan, ...
113	Bellside ironstone, ...	Bellside, Greenhill, ...
125	Kiltongue coal, ...	Coatbridge, Airdrie, Kilgarth, Rochsoles. This Redding, and the "First coal" worked at S parts of the Airdrie district, there is a sea inches of blackband ironstone, and 3 inches Kiltongue coal contains an ironstone called been used for the manufacture of Paraffin.
138	Drumgray coal, ...	Airdrie, Coatbridge, Shotts. Same as the Co coal" of Shotts. There is sometimes a sea in position with the lower Coxrod of Slama
165 180 188	First, Second, } Slaty bands of ironstone, .. Third,	The first seam is that found at Garbethill, Toc at Crofthead and Armadale. The third at In the second position of this ironstone profitably used in the manufacture of Paraf
350	Caulm limestone, ... [*Sandstone and Shale.*]	At Garnkirk, Bedlay. This seam is the termin responds in position with the Janet Peat li limestone of Muirkirk. The upper series coals found in the Kilmarnock, Dalmelling lingshire; Armadale in Linlithgowshire; and also with the Alloa, Kennet, and Clack
		The lower series of Minerals begins with the Po
440	Cowglen limestone, ...	Possil, Cowglen near Pollokshaws, Bishopbrigg
486	Upper coal, ...	These seams correspond in position with the Denny, Kinniel, ...
490	Upper ironstone, 1st, ...	
509	Main coal, ...	
572	Lower Possil ironstone, 2d, ...	
515	Gas coal, ...	Lesmahagow, Govan, Wilsontown, ...
540	California clay band, ...	Bonhill, Knightswood, Gartnavel, Skaterig, ...
543	Upper Garscadden ironstone, ...	Garscadden, Knightswood, ...
553	Lower do. do., ...	Do. do., ...
557	Garibaldi clay band, ...	Bonhill, Kelvinside, ...
588	Govan ironstone, ...	Johnstone, Linewood, Banton, ...
735	Crossbasket clay bands, ...	At Crossbasket. ....
745	Hurlet coal and limestone, ...	Hurlet, Campsie. In connection with this coal much sulphur, however, to be of much valu

It is from the positions lying between the Cowglen limestone and the Hurlet coal that the larges The Comrie, Kennedder, Izievnar, Oakley, and Cowdenbeath ironstones in Fifeshire; the Kinnie in Dumbartonshire; the Penstone, Dolphingston, and Tranent ironstones in East Lothian; th of the positions above described, and under the Possil or Cowglen limestone.—The total thick of workable ironstone may be estimated at 72,100,000 tons, of common coal at 424,621,000 tons, till the Old Red is reached.

STATEMENT,

ls, with some of the principal corresponding positions in other Counties of Scotland.

RE THE SEAMS ARE FOUND AND WORKED.	Still to work—Tons.	Inches.
....		12
n, ...	12,110,400	40
n, Baillieston, Wishaw, Holytown, Motherwell, Wishaw, ...	146,107,300	96
Coatbridge, Wishaw, Motherwell, Baillieston, Drumpeller, and 80 feet at Bredisholm, Springhill, Souterhouse, Millfield. The in the Glasgow and Rutherglen district, ...	64,706,000	108
useless coal, only worked for engine and pitman's fires,..........	2,800,000	20
Motherwell, Holytown, Rutherglen. This is the best blast fur- many years exclusively used in the manufacture of pig iron	72,840,100	40
e Wee coal of Cardowan, Airdrie, and Coatbridge, and the Virgin on districts; a good household coal, ...	7,100,000	36
his is the most valuable of Scotch blackbands. It has not been immediately surrounding Airdrie and Coatbridge. It thins out chsoles on the north, Arden on the east, and Rosehall on the south,	1,492,000	16
f inferior quality, and abandoned, ...		20
d and Legbrannock, ...	335,000	5
d, Coatbridge. The Benbar coal of Shotts, and Lady Grange	36,518,000	30
....	938,000	7
. This seam is the same as the splint coal of Slamannan and ced at Shotts Iron Works. About 8 feet above this seam, in some a seam called the "Kiltongue Mussels," composed of about 6 inches of the finest gas coal.—At Calderbraes, near Airdrie, the e called the Calderbraes ironstone, and a shale which has recently raffin.	70,580,000	60
Calderbraes ironstone,........................	4,721,000	
the Coxrod coal of Slamannan and Redding, and the "Second es a seam below this, called the Lower Drumgray, corresponding Slamannan and the "Third coal" of Shotts, ...	14,723,000	24
ill, Todbuchts, Cameron Glen, and Arden. The second is found ird at Goodaskaiel. onstone at Bathgate, lies the Torbanehill coal, which has been so of Paraffin Oil,........................	33,957,000 of 2d	12; 96 to 120
termination of the upper series of coals and ironstones. It cor- Peat limestone of Comrie, in Fifeshire, and with the Bluetower series of coals in Lanarkshire correspond in position with the nellington, and Lugar fields, in Ayrshire; Grangemouth in Stir- shire; and with the upper seams of the Edinburgh coal field; l Clackmannan coal fields, ...		78
the Possil or Cowglen limestone—		
opbriggs, Kilsyth..		
th the seams found at Possil, Keppoch, Bishopbriggs, Kilsyth, ...	18,067,000 of 1st 300,000 of 2d	12 4 to 12
....	2,481,700	12
rig, ...	2,000,000	
....	4,000,000	
....	2,000,000	
....	12,571,400	
his coal there is sometimes a band of ironstone, containing too ch value.		

e largest quantity of ironstone is taken for the making of pig iron in Scotland.
Kinniel and Balbardie ironstones in Linlithgowshire; the Croy, Banton, and Kilsyth ironstones
iian; the black band in Mid-Lothian; the clay band ironstones of Muirkirk, all lie in one or other
l thickness of the coal series, with the Ballagan beds, cannot be less than 6000 feet. The quantity
)0 tons, and of gas coal at 42,000,000 tons. The chief stratum below the Hurlet coal is sandstone,

Printed in the United States
By Bookmasters